This book has been printed digitally and produced in a standard specification in order to ensure its continuing availability

OXFORD
UNIVERSITY PRESS

Great Clarendon Street, Oxford OX2 6DP

Oxford University Press is a department of the University of Oxford.
It furthers the University's objective of excellence in research, scholarship,
and education by publishing worldwide in

Oxford New York

Auckland Cape Town Dar es Salaam Hong Kong Karachi
Kuala Lumpur Madrid Melbourne Mexico City Nairobi
New Delhi Shanghai Taipei Toronto
With offices in
Argentina Austria Brazil Chile Czech Republic France Greece
Guatemala Hungary Italy Japan South Korea Poland Portugal
Singapore Switzerland Thailand Turkey Ukraine Vietnam

Oxford is a registered trade mark of Oxford University Press
in the UK and in certain other countries

Published in the United States
by Oxford University Press Inc., New York

Reprinted 2009

ISBN 978-0-19-829147-3

Restructuring of Arms Production in Western Europe

Edited by
Michael Brzoska and Peter Lock

OXFORD UNIVERSITY PRESS

This book has been printed digitally and produced in a standard specification in order to ensure its continuing availability

OXFORD
UNIVERSITY PRESS

Great Clarendon Street, Oxford OX2 6DP

Oxford University Press is a department of the University of Oxford.
It furthers the University's objective of excellence in research, scholarship,
and education by publishing worldwide in

Oxford New York

Auckland Cape Town Dar es Salaam Hong Kong Karachi
Kuala Lumpur Madrid Melbourne Mexico City Nairobi
New Delhi Shanghai Taipei Toronto

With offices in

Argentina Austria Brazil Chile Czech Republic France Greece
Guatemala Hungary Italy Japan South Korea Poland Portugal
Singapore Switzerland Thailand Turkey Ukraine Vietnam

Oxford is a registered trade mark of Oxford University Press
in the UK and in certain other countries

Published in the United States
by Oxford University Press Inc., New York

Reprinted 2009

ISBN 978-0-19-829147-3

Contents

Appendices

Preface

Restructuring of arms production is often associated with two different processes: on the one hand, gradual restricting of military production and transformation of relevant industries for the needs of civilian production (conversion), and on the other hand, introduction of new technologies and new types of production. The two run parallel, and reflect a new political and military reality. One thing is for sure: arms industries world-wide will have to adapt to a different environment. Government expenditures on armaments will probably continue to decrease. In addition, a general trend towards more open market economics and less concern with a narrow concept of military security will weaken that traditionally close relationship between companies and national governments that US President Dwight D. Eisenhower called the 'Military–Industrial Complex' in his farewell speech in January 1961.

For arms industries in Western Europe, elements of a new environment have been in the offing for a couple of years. A greater willingness by West European governments to take change in the Soviet Union (and the process of its disintegration) seriously, problems in keeping abreast of the ever-increasing speed of technological innovation and the political decision to create a single market in the European Community combined to force industries and governments to act. Arms industries were restructured both within individual states and on the West European level. As all this happened in open societies, there is no lack of information about what was intended and what happened.

The West European case is something of a model case in restructuring of arms industries. Of course, compared with other areas of the world, for instance the former Warsaw Treaty Organization member states, restructuring occurs in a benign economic environment. However, the difficulties in Western Europe—arising out of the peculiar and intimate mixture of economics, technology and politics that is characteristic of arms production world-wide—are instructive beyond that region.

This compendium of case studies on various aspects of the restructuring of West European arms industries is the companion volume to the SIPRI research report *West European Arms Production: Structural Changes in the New Political Environment,* authored by Ian Anthony, Agnès Courades Allebeck and Herbert Wulf and published in 1990. The report contains basic data on the arms industry generated at SIPRI, as well as a projection of the future of the industry, including an analysis of the effects of the Treaty on Conventional Armed Forces in Europe (CFE). An abridged, updated version of this report appears as chapter 2 in this volume, and the most recent SIPRI data appear in appendices A and B.

Together, these two publications elucidate the fact that changes in arms production require public scrutiny and political guidance. Employment in arms production in Western Europe will decrease substantially. This will create pressures to continue to produce weapons without proper military rationale and for exports to Third World regions that have already acquired an over-abundance of weapons. The arms industry is becoming more international, making control more difficult. If governments do not internationalize their efforts at control in parallel, there is a danger that after over 40 years of national military–industrial complexes we will come back full cycle to the world of largely free-roaming arms manufacturers in the period between the middle of the 19th century and the 1930s. This danger is now possibly greater east of Western Europe, not the least because governments in Western Europe have begun to discuss co-ordination. There is also a need for governments to come together to co-ordinate the future demand for weapons. A fundamental revision of force planning is required—both on the European scene, if governments are to fulfil their pledge to continue conventional arms control beyond the present CFE Treaty (CFE II—after the Summit Helsinki CSCE Follow-up Meeting in 1992), and in the conflict spots of the Third World, where agreements on regional security, including agreed levels of arms, offer the best hopes for a more peaceful future. The time is ripe for us to change our way of thinking and to include the future of arms production on the agenda of international negotiations on arms limitation and the deployment of new systems.

Adam Daniel Rotfeld
Director of SIPRI
December 1991

Acknowledgements

The origins of this book go back to a research project on arms production in Western Europe carried out in the late 1970s by Dan Smith and Clare Mundy in London and by Herbert Wulf and the undersigned in Hamburg. Since then, we have had the benefit of discussing the subjects of this book on a great number of occasions with a large and diverse group of people.

Earlier versions of the chapters in this book were originally presented on 6–8 June 1990 at a seminar at House Rissen in Hamburg. A generous grant from the Volkswagen Foundation allowed us to assemble a distinguished group of academic experts. Papers were later revised in light of the comments made at the seminar. Not all of the papers presented at the seminar are included in this book. We decided to stick to a narrow thematic focus. Some interesting papers of high quality on issues such as the future European security order and on conversion of arms industries from military to civilian production were ungracefully sacrificed. We are indebted to Alberto Castagnola, Francois Chenais, Agnès Courades Allebeck, Klaus Engelhardt, Sami Faltas, Jean-Paul Hébert, Hartwig Hummel, Helmut Maneval, Arcadi Olivares i Boadilla, Steven Schofield, Kurt Tudyka and Helena Tuomi for their contributions during the seminar. Of course, we alone are responsible for errors and omissions.

Thanks are due to Henning Schierwater's staff at House Rissen who provided us with a stimulating working atmosphere. Marc Vogel helped with practical arrangements. It is only due to the indefatigable work of Paul Claesson at SIPRI that this book has become an English-language publication; in fact, he did more than that, transforming manuscripts into a coherent and readable book. Last but not least we are grateful to our patron at SIPRI, Herbert Wulf, leader of the arms trade and production project; SIPRI's former Director Walther Stützle; and SIPRI's current Director, Adam Daniel Rotfeld, for making this publication possible.

Michael Brzoska and Peter Lock
December 1991

Abbreviations and acronyms

ACV	Armoured combat vehicle
AEW	Airborne Early Warning
ASEAN	Association of South-East Asian Nations
ASRAAM	Advanced short-range air-to-air Missile
ATF	Advanced Tactical Fighter
AWACS	Airborne warning and control system
BAe	British Aerospace
Benelux	Belgium–the Netherlands–Luxembourg
CAPS	Conventional Armaments Planning System
CATT	Conventional Arms Transfer Talks
CBO	Congressional Budget Office
CEPA	Common European Priority Area
CFE	Conventional Armed Forces in Europe
COCOM	*Co*-ordinating *Com*mittee on Multilateral Export Controls
CRS	Congressional Research Service
CNAD	Conference of National Armaments Directors
CSCE	Conference on Security and Co-operation in Europe
CTP	Cooperative Technology Project
DARPA	Defense Advanced Research Projects Agency
DASA	Deutsche Aerospace
DDIs	(Countries with) developing defence industries
DESO	Defence Export Services Organisation
DIDA	Defence Industry Development Agency
DGA	Délégation Générale pour L'Armement
DOD	Department of Defense
DV	Delivery verification
EC	European Community
EDEM	European Defence Equipment Market
EDIG	European Defence Industry Group
EDIS	European Defence Industry Study Group
EFA	European Fighter Aircraft
EFTA	European Free Trade Association
ETP	European Technology Plan
EUCLID	European Co-operative Long-term Initiative for Defence
Eureka	European Research Co-operation Action
FMS	Foreign Military Sales
GDP	Gross domestic product
GNP	Gross national product
IEPG	Independent European Programme Group

IGC	Intergovernmental Conference
LDCs	Less developed countries
LDDIs	(Countries with) less developed defence industries
MBB	Messerschmitt–Bölkow–Blohm
MIC	Military–industrial complex
MILREL	Military-related
MOD	Ministry of Defence
MOU	Memorandum of Understanding
MSOW	Modular Stand-Off Weapon
NAD	National Armament Director
NATO	North Atlantic Treaty Organization
OECD	Organization for Economic Co-operation and Development
R&D	Research and development
R&T	Research and technology
RITA	Réseau Intégré des Transmissions Automatiques
RTP	Research and Technology Programme
SDI	Strategic Defense Initiative
UKADGE	United Kingdom Air Defence Ground Environment
UN	United Nations
WEU	Western European Union
WTO	Warsaw Treaty Organization (Warsaw Pact)

Conventions in the tables

. .	Data not available or not applicable
–	Nil or a negligible figure
m.	million
b.	billion (thousand million)
$	US $, unless otherwise indicated

Part I
Introduction

1. Restructuring of arms production in Western Europe: introduction

Michael Brzoska and Peter Lock

I. The changing foundations of arms production

In the late 1980s a predominant pillar of the post-war global order was shattered. Instead of the East–West conflict as such, the unwinding of this confrontation began to dominate political thinking and action. Many consequences of this dramatic change in international relations have begun to surface as the process continues to gain speed. One is the effect on industry of reduced military spending. Since the military focus of the East–West conflict was in Europe, it is there that the most pronounced consequences are to be found.

The termination of the East–West conflict as a military contest, important though it may be, is but one of a number of factors that will affect the provision of arms in the 1990s. Considerable academic, industrial and political thinking, discussion and activity are devoted to charting and analysing these factors. Some apply universally, and some are specific to certain regions. While there are good reasons to study changes in arms production in other regions—the Soviet Union, Eastern Europe and the Third World—arms production in Western Europe is of special interest. This is not only because it is the focus of recent changes in East–West relations, as demonstrated by the success of efforts to reduce substantially the levels of conventional armaments, but also because of its relation to the integration of Europe, or rather, at least for the time being, of Western Europe. The integration of Western Europe, a term used here in its widest possible sense, has gained much momentum following the 1986 decision of the members of the European Community (EC) to create a true common market. There is a trend both to expand the mandate of the EC—to allow it to become a political, and possibly military, as well as an economic union—and to expand the Community as such. In light of the developments in East–West relations that have taken place since the autumn of 1989, these two trends are not as difficult to reconcile as they were often seen to be in the past. Arms production, on the borderline between economics and security policy, is both embedded in the general dynamics of integration and a factor shaping its form.

In addition to international political factors, a number of other factors have increased the pressure to change the established patterns of arms production and procurement in Western Europe in recent years. These include the drive of technology towards more complex and expensive weapon systems, changing relations between military and civilian technology, the limited regulation of arms production and a general trend among governments in advanced industrial countries to reduce the economic role of the state. These factors alone might have sufficed to lead to a restructuring of arms production in Western Europe. With both East–West *détente* and European integration on the agenda, the West European arms industry is at a crucial stage. Developments and decisions have already taken place that have thoroughly disrupted established patterns of military–industrial relations. Considering the extent of changes occurring in the framework of arms production and the importance that has often been attached to arms production in terms of employment, technology, national security and foreign policy, it is safe to assume that more upheavals are in the making. Decisions will have to be made that affect not only the jobs of hundreds of thousands of workers in arms production but also national and international government policies in many areas, including technology, regional economics, procurement of weapons, military alliances and international integration.

This introductory chapter is a brief overview of the changing foundations of arms production with special emphasis on Western Europe. Aspects of technology are presented in section II; the limits of regulation in section III; and the expansion of markets in section IV. Observations of a more general nature that can help to provide a basis for discussing dilemmas, policies and possible futures are presented in section V. The issues raised here are discussed in more detail in the chapters that follow, mainly on a country-by-country basis. Relevant literature containing more profound argumentation and data is collected in a select bibliography at the end of the book.

II. Aspects of technology

Technology drive

The pace of technological development has, at least since the industrial revolution, probably grown progressively from decade to decade. Still, for the current observer, recent advances seem to be especially breathtaking—the greatly increased use of computers, and electronics in general, is often interpreted as a 'revolution' in itself, the information revolution. However, while the most spectacular advances may have taken place in the area of electronics, these are only part of a general increase in knowledge.

As a matter of course, weapon systems have absorbed even the most recent technologies, sometimes before they were used for civilian purposes. Many new technologies have emerged in the course of the development of weapon systems, such as highly integrated chips and heavy-duty ceramics. The most important effect of the technological dynamics is that the research and development (R&D) costs for any new generation of weapon types continue to increase dramatically, by factors of two to five.

The ever larger R&D resources required to fund new weapon systems raise the price of weapon systems in at least three ways. First, with each particular weapon system, a larger sum of R&D has to be recovered. Second, without proportional increases in weapon budgets, R&D absorbs a growing proportion of the available resources. The R&D costs thus are spread over a smaller number of weapons. Third, this decrease in numbers causes further cumulative price increases because 'economies of scale' are of considerable magnitude in the arms industry, particularly since there are decreasing 'learning costs' associated with production. The cost of the first item produced is much higher than that of the 100th, which in turn is higher than that of the 1000th.

Sectoral shifts

The information revolution has other effects. For instance, it leads to shifts in the importance of subsectors in arms production. Traditional subsectors, such as shipbuilding and vehicle or aircraft construction, are now less important than electronics. Another activity gaining relative importance is the integration of subsystems, including electronic components, via hardware and software.

The growth of the electronics industries was nourished by military subsidies in the 1950s and 1960s. In the 1970s and 1980s, however, they outgrew military markets. Their outstanding dynamism derived mostly from civilian products covering a broad range of applications. In general, the companies benefiting from sectoral shifts tend to be civilian-oriented while those most likely to lose are those that are heavily dependent on military orders.

Predominance of civilian technology

The distinction between civilian and military technology, always arbitrary and in practice difficult to draw, tends to vanish. It makes more sense to talk about a joint 'technology pool' to which research from both the civilian and military sectors contributes (see chapter 6)—although the secrecy of military research impedes a full sharing of advances.

Nowadays new technological advances in innovative sectors like electronics tend to be stimulated by civilian markets. Components developed for civilian purposes are more often later integrated into weapon systems than the other way around. Companies producing in related dynamic subsectors of arms production have to be aware of civilian technological advances. In fact, they are likely to be predominantly oriented towards civilian markets, with military production as a sometimes quite profitable 'spin-off'. Intimate relationships between the military, procurement agencies and the arms producing industry, for which the term military–industrial complex (MIC) has often been used since it was first mentioned by US President Dwight Eisenhower in 1961, are counterproductive in this kind of setting. Competition requires a quite different corporate environment from that resulting from cost-plus contracts.

New competitors

The changing emphasis of military and civilian technology has also changed the pattern of competition in the arms industry. It shifts competition among specialized arms producers into the arena of world-wide industrial competition. This frame is often seen to be dominated by a triangular competition between US, West European and Japanese industries. While competition with US companies is nothing new for arms producing companies in Western Europe, Japanese industry has not yet played a role. The low profile of Japanese companies in arms production has not only prevented them from becoming major arms exporters, it has also meant that they did not aim to be on the forefront of technological advance in the military field. Nevertheless, Japanese dual-use components are first-rate and are applied in many weapon systems.

III. The limits of regulation

There are many ways to manage politically the constant drive for technological innovation, with its consequences of increasing unit costs and sectoral shifts. One way is to ignore the trends, keeping production at a given level of technology. There may be no obvious necessity all the time to incorporate the latest technological advances into weapon systems, but there is an empirically established general tendency among soldiers, bureaucrats and politicians in charge of procurement to favour the most sophisticated technologies.

No single national armament market in a West European country is sufficiently large to support a full range of domestic arms producers working at the most advanced level of technology. One way that has been tried to cut costs and have a domestic arms industry simultaneously is to produce jointly with other partners. During the past few decades co-operation in development and production of weapons has become widespread in Western Europe.

The limits of co-production

Co-production ventures are usually the results of economic necessity, not the preferred choices of either companies involved in arms production or governments. Companies want to do as much business as possible; they are also suspicious of having to share technology with possible competitors. Governments tend to be unwilling to support 'sell-outs' of either jobs or technology. Because of alliance politics or simple compensation for favours in another area, they are seldom willing to pay more for co-production than national production would cost, as in the case of the Franco-German Tigre anti-tank helicopter project.

Co-production allows a sharing of the R&D costs and makes production runs longer. While this tends to lower unit weapon costs, other factors—translation, co-ordination, management, control and regulation alone—increase costs. More important, costs depend strongly on regulation. Work shares have to be divided up, usually in proportion to financial contributions to the project. Production has to be constantly controlled to ensure that *juste retour* between financing and production shares is assured. In extreme cases, the reservation of national shares leads to the 'incompetence principle'.[1] Particularly a less competent company in a junior role can learn more from participation in a co-production project with a technological leader from another country than a competent one.

The exact financial balance of co-production is therefore a matter of debate, or rather the point of comparison. Purely national projects would in most cases result in higher unit costs, while production for several customers without the industrial and bureaucratic requirements of co-production would be cheaper.[2]

The question of free markets

In an environment of protected national markets, administered co-production may be the best policy achievable. If protection is not paramount, however, its limitations become obvious. A free market for arms and no government interference in production would seem to be optimal for the industry. However, governments often cite several reasons for regulation of domestic arms markets.

The most important one is that without government interference there is little chance that companies will invest in highly risky projects at the edge of technology. Governments subsidize to get firms involved. Sometimes governments have even subsidized several companies within the same fields of technology to keep competition alive. An example is West German submarine production, where government subsidies are shared by HDW and Thyssen Nordseewerke. This is obviously a very expensive approach to keep some competition alive.

Another reason is to ensure political control over arms producing enterprises. The image of 'merchants of death' still looms large in the general public. Until recently, left-wing political programmes regularly included the nationalization of arms industries. Experience has shown, however, that state ownership does not totally guarantee legal behaviour. For instance, the state-owned Austrian firm of Hirtenberger was involved in illegal sales to Iraq and Iran. The state-owned Spanish firm of Santa Barbará was involved in illegal sales to Argentina.

A third reason is the potential to exploit the ascribed economic and technological benefits of arms production. For instance, arms procurement can be directed to depressed areas. Although such action is economically suboptimal, it may be the 'second best' option, since depressed areas have to be subsidized in any case. However, current economic thinking tends to question this rationale, as subsidies in general are considered suboptimal. Given the volatility of the market, specific subsidies to arms industries are even more questionable.

[1] Mattingly, M., 'Conventional armaments co-operation: putting the cart before the horse', *NATO Review*, vol. 35, no. 5 (Oct. 1987), pp. 26–30.

[2] Compare the differing assessments in Moravcik, A., 'The European armaments industry at the crossroads', *Survival*, vol. 32, no. 1 (1990), pp. 65–85; and Bittleston, M., *Cooperation or Competition? Defence Procurement Options for the 1990s*, Adelphi Paper no. 250 (International Institute for Strategic Studies: London, spring 1990).

All of these arguments assume that it is in the national interest to have a broad-based domestic arms industry, even if it is costly. The regulation of domestic arms markets results from national security thinking that puts a high value on autarchy and autonomy. It runs counter to the general economic philosophy in at least the major West European countries that an economy produces best results if left alone.

The British Government was the first in Western Europe to scrap the old style national security approach to arms procurement. In the 1980s the British Government introduced guidelines for procurement policy that claimed to be largely free from such considerations. The 'Levene' reforms of 1984 introduced more market elements into arms procurement than ever before, in the UK and elsewhere in Western Europe. Although not free from interference, as demonstrated when the Government did not give a go-ahead to GEC to take over Plessey in 1986, the British procurement reforms cleared a new path—away from regulations and towards market policies.

These reforms became important because they coincided with a general reorientation of industrial policy throughout Western Europe. Privatization and market orientation were much more in the line of prevalent economic and political thinking than in the decades before. However, uncertainty remained in other countries, as in the UK, as to whether more competition in the same market was not a one-dimensional approach to a multifaceted problem. The UK is in a privileged position because it has a number of producers that can compete on the world level. However, in most other countries, as in the UK, a radical market approach would lead to large cuts in employment and the demise of whole sectors of arms production. It would most probably be US companies that would gain most from such a strategy since they operate in by far the largest market for armaments.

IV. Expansion of markets

The above discussion can be summed up in one proposition: national markets are too small. The rapid evolution of military technology seems to make it constantly smaller. Regulation or free market policies are only—sharply differing—stop-gap measures to avoid a full appreciation of the problem. The obvious solution is to expand the marketplace without letting superior competitors compete on fully equal terms (see figure 1.1).

The limits of joint procurement

It is of course possible to give up national procurement and to co-ordinate it on a multilateral scale without co-ordinating production at the same time. This has been proposed on several occasions by various actors ever since NATO came to function. Despite appealing economic advantages these proposals never made much headway. What came about instead was a multiplication of institutions

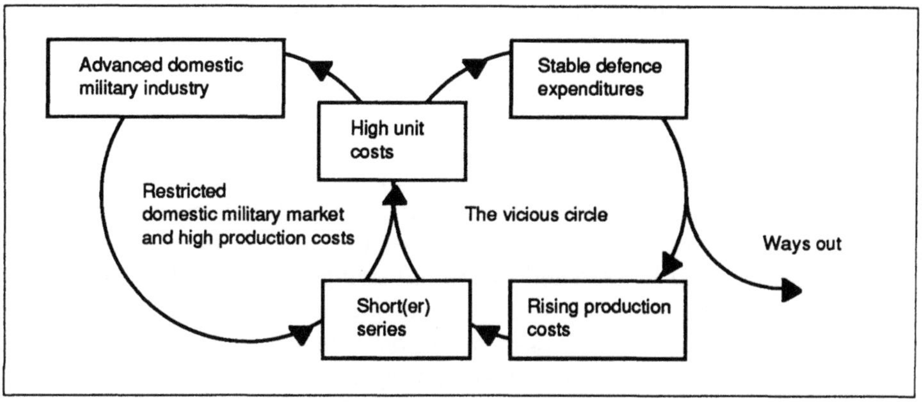

Figure 1.1. The vicious circle of an expanding arms market

Source: The figure was developed by Björn Hagelin.

and of co-ordination forums, but only few cases of joint procurement. In fact, the NATO co-ordinated procurement policy has not resulted in a single major weapon project since the Italian G-91 fighter aircraft was adopted as the 'NATO fighter' in the 1950s. All the same, much energy has been invested in promoting common procurement .

Why this preponderance of national procurement? An important reason was certainly the resistance of nation-states to give up sovereignty, namely, national decision-making over weapon systems, armed forces and arms industrial policies. There may be more progress towards military integration in Western Europe in the future. However, this is not likely to come from the procurement side but rather in conjunction with the accelerating advances in West European economic integration.

The importance of the Single European Market

The creation of the Single (West) European Market effectively erodes national sovereignty in important economic policy areas. It radiates into virtually all spheres far beyond its already great economic significance. This is also true for arms production, although it is nominally excluded from the common market (see chapter 3). The importance of the Single Market can be traced on two levels: that of national governments and that of companies.

It becomes increasingly difficult for national governments to justify the special treatment of the arms production sector. Why should a national industrial policy be restricted to a sector that hardly can be delimited any longer? Also, on the procurement side it becomes awkward to have special treatment for some items while others also important for national security and close to weapons fall under guidelines applicable throughout the EC. Most important, however: an argumentation based on domestically defined national security considerations is counter to the general trend of closer integration, planned to result in a political union.

Company activity: co-operation and mergers

For companies the consequences of the Single Market are obvious. They have to strengthen their competitiveness, through rationalization, co-operation and mergers. The jockeying for positions started even before the signing of the Single European Act in 1986 and was in full swing in the late 1980s and early 1990s. Companies with interests in arms production could not make an exception. Their civilian businesses had to be reshaped anyway, and it makes no corporate sense to keep arms production separate. Although arms production seldom dominated company decision-making, its corporate structures were substantially changed through a large number of mergers, acquisitions and co-operation agreements. The fast pace at which arms production units were shed and bought, merged and split was unusual if compared to past behaviour in the arms production sector, although not if compared to what happened in many civilian sectors.

Industrial changes came about both within national boundaries and beyond them. An example of the first type was the widely discussed merger of Daimler Benz and MBB in the Federal Republic of Germany (see chapter 10). There were also, on a much larger scale than ever before, sales and acquisition, co-operation ventures and mergers beyond national boundaries, many explicitly within West Europe. Thomson of France, for instance, acquired production bases in a number of West European countries (see chapter 8). However, world-wide strategies induced companies based in Western Europe to look beyond Europe. Thus, a number of West European companies bought US interests or negotiated wide-ranging co-operation measures with US companies. Recently Daimler Benz announced a close link with Mitsubishi of Japan.

In the dynamics of change in the arms production sector, the center of gravity shifted in the late 1980s towards the arms producers, as the most dynamic companies that began to cross national boundaries and to seek larger markets.

The prevailing trends are towards a more integrated arms industry, mainly on national levels, but also on an international level. Given the economic and political links in Western Europe, integration occurs in the—largely overlapping—framework of the EC and European NATO (see chapter 3). Its most dynamic elements at the moment are the arms producing companies.

V. Prospects of restructuring

The fundamental changes in relations between East and West, and among West European countries, have combined with the constant drive of technology to trigger a restructuring of arms production in Western Europe that may be fundamental. Much is in flux and as yet difficult to predict, but the circumstances of uncertainty make the discussion more interesting and potentially more rewarding. Of course, solid analysis should not give way to speculation.

A number of policy areas are affected by changes in arms production or can be expected to be affected if current trends continue. Obviously there is still much uncertainty about the exact course of events, not the least because of the important effects that future political decisions, such as those pertaining to European integration, will have on, for instance, the level of armaments in Europe. The questions below focus on the future implications of restructuring in three policy areas where changes might have important effects, and where there is much room for alternative political decision making. It is with these questions in mind that the authors of the following chapters were invited to contribute their expertise.

Military and disarmament policies

One factor behind the prevalent 'buy national' strategy has been the wish to keep an independent military force and posture. It may be assumed—which, given the commitment to political union in Western Europe, is not obvious—that this is to remain for a while. But will it be possible with West European arms industries far more integrated? Does integration 'from below' allow for non-integrated military postures? Or will there be an additional impetus to integrate military forces as well?

With the possible demise of the East–West conflict, disarmament comes within reach. There are several ways in which governments and companies can react to reduced procurement budgets. They range from an expansion of arms exports to the laying off of workers. Can larger entities better cope with reduced demand? Will international integration introduce additional complications in policies aimed at reducing the negative impact of decreases in production? Is conversion of arms production capacities to civilian production easier or more difficult than integration than other strategies?

Technology and economic policy

As outlined above, arms production is, among other factors, the result of considerations of potential economic and technological objectives. More integration reduces the possibility to meet these objectives on a national level—but in what ways? For instance, will it still be possible to use arms production as an instrument of regional and sectoral policies? At the same time the hope to save money is an important motive behind the spur to transnational restructuring. Will restructuring lead to reduced prices of weapon systems? Or will it recreate price rigidity on a higher level? Will companies be more willing to commit finances to military R&D than they have in the past?

International relations

It is still difficult to predict how protective the West European arms industry will be. It cannot ignore the US market, by far the largest in the world. For the same reason, an unprotected West European arms market would in all likelihood be overrun by US companies. What does this mean for the relationship with the USA in this field? What are the likely reactions of US companies and the US Government to a protected Single European Market for armaments? How can this affect general relations between Western Europe and the USA?

As mentioned, one possible reaction to shrinking markets is the expansion of exports to Third World countries (see chapter 4). As of now there are some differences in arms export policies among West European countries.[3] Will they be eliminated and which policy will set the standard? On what level? Will Western Europe be more or less willing to join or even initiate limitations of transfers of military technology to Third World countries?

Modes of restructuring

This discussion leads back to the idea of the military–industrial complex. If arms procurement has been suboptimal, restructuring and internationalization could change it to become more efficient. On the other hand, if the MIC is just re-erected on a higher—i.e., EC—scale, it might also recreate its problems on a higher scale, that is, with respect to disarmament policies. The political consequences of restructuring are obviously not independent of the structural peculiarities of this restructuring. There are a number of such elements.

Private versus state ownership

There still is quite extensive state ownership in arms production, although less so in northern than in southern Europe. In addition, governments are trying to withdraw from direct management to positions of simple shareholding. Does state ownership change perspectives on restructuring? Does restructuring change perspectives on state ownership?

Regulation of markets

The arms market is still characterized by extensive regulation. Lured by wider markets, arms producing companies may be willing to go beyond national boundaries, but they are also interested in keeping the benefits of domestic protection. Governments may not be willing to give up the strategic aim of having and regulating a national industry. Both companies and governments seem to be looking for ways to combine markets and regulation—but to whose benefit, and how?

[3] See Anthony, I. (ed.), SIPRI, *Arms Export Regulations* (Oxford University Press: Oxford, 1991).

Part II
The changing environment

Demand and supply

At the onset of the 1990s, companies are the more dynamic element in restructuring. Do existing institutions support this dynamism? What kind of institutional changes would be in compliance with that type of dynamic? Will the EC play a larger role in shaping the framework for arms production? Companies will not necessarily remain the focus of restructuring. The initiative may well shift back to procurement institutions. Is it not logical to have more common procurement institutions to cope with more integrated and hence oligopolistic companies? If so, how can parliamentary control be assured?

Co-production, consortia and free markets

Integration has occurred in various ways, ranging from outright take-overs to agreements of co-operation. Will a more stable pattern emerge? Is co-operation to last, or will it prove to be a phase before merger? What will be the role of co-production in the future? Is a 'free market' in arms production possible? Is it desirable? Will the 'free market' lead to even more powerful concentration?

VI. What to expect in this book

The chapters in this book are factual accounts of what has happened in arms production sectors in a number of countries in Western Europe. There are chapters on the major arms producing countries, where the dynamics of restructuring have been most pronounced. Since there is considerable literature available describing the structures of arms industries in the FRG, France and the UK, these chapters focus on the recent changes that have occurred or those expected to occur. As the arms industries in other countries are not so well documented, the chapters on these countries contain more detailed descriptions of the industries as such.

Other chapters describe some of the processes which are directly related to the restructuring of arms production in Western Europe: efforts to provide institutional frameworks for the restructuring of arms industries in Western Europe, developments in the world arms export markets, and the reaction to changes in international relations by the US arms industry. This selection is somewhat arbitrary; there is no chapter on the larger processes shaping the future of arms production in Western Europe, such as the future of the EC, negotiations on the reduction of conventional weapons in Europe or the future of NATO. Relevant as these themes are to the subject at hand, they are too important to be treated as side-issues. All the chapters reflect the authors' familiarity with the underlying political issues of the future, but the focus in this book is on recent changes and the near-term future of arms industries.

2. The economics of the West European arms industry

Ian Anthony and Herbert Wulf[1]

I. The size of the arms industry

Of the three major centres of arms production in the world, the USA, the USSR and Western Europe, the West European industrial base is the smallest. The US Government spends more than twice as much on the procurement of major weapon systems than all European NATO countries together, including France. Similarly in exports, the USA still ranks higher than the West European countries combined. According to SIPRI statistics the USA exported major weapon systems worth $53.8 billion for the period 1986 to 1990, second only to the USSR. In 1990 the USA accounted for 40 per cent of the world total of major weapon exports. The 12 member countries of the European Community (EC) combined accounted for just half of this—20 per cent of the total value of major conventional weapons delivered in 1990. Within that total, the share of deliveries accounted for by the Federal Republic of Germany, France and the United Kingdom was 90 per cent.

The major West European companies

Corporations from nine different countries are represented in the list of the 50 largest West European arms producing companies (see appendix A). Table 2.1 indicates the high concentration of companies from the UK (15), France (12) and the FRG (10): these 37 companies account for over 75 per cent of the arms sales of the 50 companies listed. This is a reflection of the fact that the UK, France and the FRG account for three-quarters of NATO Europe's total spending on the procurement of major weapons, and the fact that British, French and German companies are active arms exporters.

The remainder of the list is made up of four Swedish companies, three Italian, three Swiss, and one each from Belgium, the Netherlands and Spain. Table 2.2 illustrates the degree to which arms sales by the major West European companies are concentrated.

[1] This chapter is an abridged and updated version of chapters 3, 4 and 6 of a report by Anthony, I., Courades Allebeck, A. and Wulf, H., *West European Arms Production: Structural Changes in the New Political Environment*, A SIPRI Research Report, Oct. 1990.

Table 2.1. Number of companies in the list of 50 largest West European arms producing companies, grouped by rank and country, 1989

	Rank					Total no. of companies	
Country	1–10	11–20	21–30	31–40	41–50	1989	1988
UK	3	1	4	5	2	15	16
France	5	3	–	2	2	12	11
FRG	1	1	2	2	4	10	9
Italy	1	2	–	–	–	3	3
Spain	–	1	–	–	–	1	1
Switzerland	–	1	1	–	1	3	3
Sweden	–	1	2	1	–	4	4
Netherlands	–	–	1	–	–	1	2
Belgium	–	–	–	–	1	1	1

Source: Appendix A.

The largest producer, British Aerospace, accounts for 11 per cent of the arms sales of the 50 largest companies. The 5 largest companies account for 38 per cent, and the 10 largest companies for 56 per cent. These concentration ratios indicate that the arms business in Europe is primarily the activity of a very small number of corporations. Moreover, as discussed below, recent mergers and acquisitions have concentrated arms production further. The extent to which these companies depend on arms production is particularly important in a period—as at present—when contractors expect a further cut in orders. Most companies have at least some interest outside the arms business and only a few corporations among the 50 largest European arms producing companies are highly dependent on arms production (see table 2.3). Some of the 50 largest arms producing companies are highly diversified, with only a fraction of their sales in the arms sector and 14 have arms sales of 10 per cent or less of their total sales. However, a number of holding companies with diversified business activities have subsidiaries that are highly dependent on arms sales.

Table 2.2. Arms sales as a share of total sales for companies in the list of the 50 largest arms producing companies

Company rank groups	Percentage share of arms sales of the 50 largest companies
1	11
1–5	38
1–10	56
1–25	82

Source: Appendix A.

Table 2.3. Arms production-dependent and diversified companies among the 50 largest arms producing companies in Western Europe, 1989

Rank	Company	Country	Arms sales as share of total sales (%)	Arms sales (US$ m.)
Arms production-dependent companies				
4	DCN	France	*100*	3 630
17	GIAT	France	*97*	1 020
21	VSEL Consortium	UK	*99*	870
26	Eidgen. Rüstungsbetriebe	Switzerland	*94*	550
36	Devonport Management	UK	*98*	410
Diversified companies				
3	Daimler Benz	FRG	*10*	4 260
6	IRI	Italy	*5*	2 230
12	INI	Spain	*8*	1 290
13	Fiat	Italy	*3*	1 280
19	Siemens	FRG	*3*	900
22	Philips	Netherlands	*3*	800
27	Thyssen	FRG	*3*	540
28	Thorn EMI	UK	*9*	540
29	Saab-Scania	Sweden	*7*	530
33	Krupp	FRG	*5*	440
39	Renault Véhicules	France	*1*	390
40	Ericsson	Sweden	*6*	360
41	Mannesmann	FRG	*3*	360
42	Hawker Siddeley	UK	*10*	350

Source: Appendix A.

Politics versus economics

As indicated in figure 2.1, even the largest arms producing companies are relatively 'small potatoes' in European economic terms. The sales of the largest companies in non-military industrial sectors are several times larger than those of the top companies in arms production. It is for this reason that the EC takes a fairly detached interest in the activities of arms manufacturing companies.[2]

The primary economic importance of these companies lies in their leading role in technology development. However, even this role has increasingly been challenged by the rapid growth of research and development (R&D) spending, especially within the civilian electronics sector. The introduction of new

[2] Eurostrategies, 'Dual-use industries in Europe', unpublished study prepared for the Commission of the European Communities, Directorate-General III (Internal Market and Industrial Affairs), Jan. 1991.

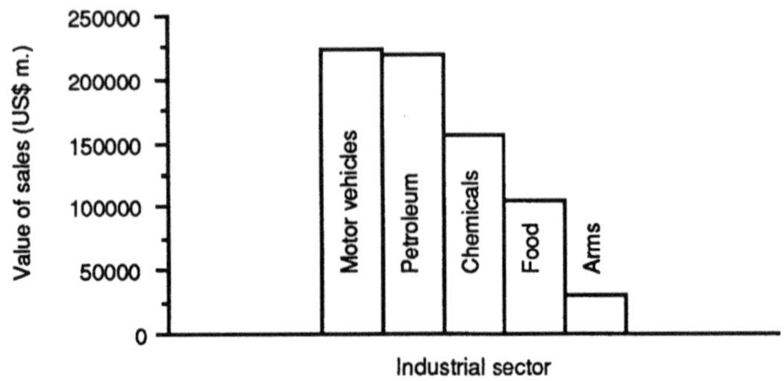

Figure 2.1. Comparative sales figures of the 10 largest European companies in each of selected industrial sectors, 1988

Sources: 'The world's biggest international corporations', *Fortune 500 1988*, 31 July 1989, pp. 45–70; SIPRI data base.

technology, especially in information technology, has become more and more a function of the commercial sector rather than military priorities.[3]

Deciding the future of the arms industry should be seen as a political problem with a limited economic dimension. There is no sound reason why people and companies making arms should be afforded special treatment by politicians, in particular when arms manufacturing is of relatively minor economic importance to the overall European industrial base. If logic were to prevail, the size and shape of the future European arms industry would be a function of the security needs of the respective arms producing countries. In reality, however, arms procurement will probably continue to respond to a range of political factors of which national economic performance will be one of the most important.

As companies experience declining arms sales—both domestically and in exports—arms industrial policies oriented towards national economic considerations should give way to policies designed to encourage rationalization—that is, a reduction in over-capacities and the duplication of design and production capabilities—throughout the West European arms industry. While governments subscribe to this principle, they have yet to find a satisfactory mechanism for giving substance to it.

Governments have a unique degree of control over the arms industry in that it is one of a very few industrial sectors where governments regulate the speed of technical progress, the volume of production, the price of goods and whether or not a company will be permitted to sell in foreign markets.[4] As indicated in

[3] See chapter 6. The changing impact of technology development on arms production is discussed in Gummett, P. and Reppy, J. (eds), *The Relations between Defence and Civil Technologies* (Kluwer Academic Publishers: Dordrecht, 1988).

[4] Hartley, K., 'The European defence market and industry', eds P. Creasey and S. May, *The European Armaments Market and Procurement Cooperation* (Macmillan: London, 1988).

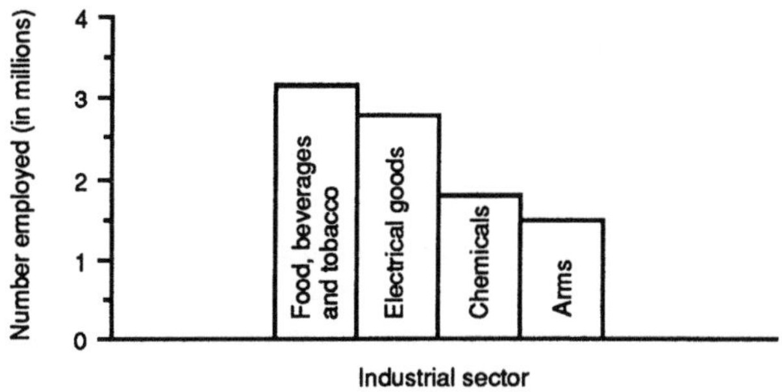

Figure 2.2. Total employment in selected West European industrial sectors, 1987

Sources: *Statistique de base de la Communauté* (Office des publications officielles des Communautés européennes: Luxembourg, 1989); Wulf, H., 'Westeuropäische Rüstungskooperation zwischen bürokratischen, industriellen und militärischen Interessen', ed. R. Seidelmann, *Auf demWeg zu einer westeuropäischen Sicherheitspolitik* (Nomos: Baden-Baden, 1989), pp. 160–63.

figure 2.2, the employment impact of arms manufacturing is greater than the financial importance. In contrast to sales figures, levels of employment in arms production are comparable in size to other industrial branches in Europe. Therefore, historically, no government has been able to resist the temptation of using arms procurement as an element of industrial and especially employment policy. There is no reason to believe that the future will be different.

Arms manufacturing companies are faced with a 'shrinking pie' in the form of declining arms procurement expenditures. The compilation of data on the 1989 arms sales[5] of the 100 largest arms producers in the OECD (Organization for Economic Co-operation and Development) countries and in the Third World reveals that the trend of reduced arms production has already been reflected in company sales. The arms sales of many companies have dropped and West European companies are no exception. Since this trend has not changed as a result of the Persian Gulf War, many of these companies will receive fewer contracts than they did in the past. Production capacities will have to be reduced. A more precise prognosis of the future of the arms industry depends on how defence ministries will distribute cuts between the various headings in the budget. The outcome is likely to be different in each of the major West European countries, reflecting their different security priorities. These are discussed for each country in detail in the chapters in part III.

[5] The term 'arms sales' is used here to describe the true value of sales by companies to domestic as well as foreign customers as derived from company reports, government documents, journals, newspapers, etc. This differs from SIPRI's usage of the terms in estimating the value of the international trade in major conventional weapons.

II. Trends

Procurement budgets and arms exports

Economic burdens from investment in the military sector have been one cause of a thorough revision of Soviet military policy. In the USA the procurement budget has already been cut in real terms by almost 16 per cent from its peak year of 1987. Further reductions of up to one-quarter by the mid-1990s are planned. The financial burden of military expenditure is a concern not only in the USSR or in the USA. Budget constraints in many West European countries have grown as a result of competing domestic economic priorities, particularly in the context of a generally more favourable East–West climate. The gradual success of arms control negotiations has had a political impact, affecting governments' readiness to allocate funds to the military. The perception of a disappearing threat and the fading of enemy images seem to be giving finance ministries more power to question military budgets. This is clearly reflected in the budget situation at the beginning of the 1990s.

Procurement expenditures by European NATO countries, which had increased in the early 1980s, began to level off in 1984 and reached their peak in 1987 with expenditure of approximately $35 billion (see appendix C). In real terms, procurement expenditure in 1990 was worth approximately the same as it had been in 1982. During the second half of the 1980s the arms industry was confronted with a stagnating domestic market in Western Europe.

Moreover, the second half of the 1980s also saw the largest arms producing West European countries facing shrinking export orders. The total value of new agreements signed with Third World countries by the four largest arms exporting countries in Western Europe—the FRG, France, Italy and the UK—fell from $5.1 billion in 1983 to $4.2 billion in 1990, according to data published by the US Congressional Research Service.[6] It is clear that the decline in export orders by European countries would have been a more visible feature if it had not been for two extraordinary peak years for British arms export orders. These peaks—in 1985 and 1988—reflect the signature of memoranda of understanding relating to massive 'packages' of British arms to be supplied to Saudi Arabia and to Malaysia, respectively. However, whether or not these will in reality lead to contracts being signed remains an open question.

With a significant number of new national and co-operative programmes in the planning and development stage, arms producing companies were still optimistic about their medium-term business prospects as late as 1989. However, in the dramatically changed international climate, and with the financial situation in many countries becoming increasingly difficult, military expenditure is less and less likely to be spared from downward adjustment. The arms industry faces an entirely new situation.

[6] Grimmett, R. F., *Conventional Arms Transfers to the Third World, 1983–90*, Congressional Research Service Report no. 91-578 F (US Library of Congress: Washington, DC, 2 Aug. 1991), p. 46.

The future of procurement

Of the European NATO countries only Turkey has planned substantial increases in the procurement of major equipment. Procurement budgets in European NATO countries have either already been cut or are likely to be cut in the 1990s. Long-term plans have been revised in several countries where security policy, force structures and procurement plans are being reconsidered. In important West European military powers, including Germany and the UK, initial modifications to the armed forces will focus on reductions in manpower levels.

The defence industry will, in future, operate in circumstances where NATO force posture has been significantly altered in response to successful disarmament initiatives. As one official describes the situation: 'limitations on a number of key equipments will lead to fewer acquisition programmes, shrinking national demands, reduced export prospects and therefore probable rises in unit costs. This will put increasing pressures on the NATO defence industries, and this may lead to further structural changes in these industries'.[7] Over the longer term it is not unrealistic to expect annual cuts in the order of at least 2–3 per cent in real terms in the procurement budgets. Deeper cuts may follow in response to further arms control agreements.

While cuts in real terms are the norm for procurement budgets, this is not the case in military R&D, which has not yet been affected to the same extent. On the contrary, military R&D is being funded in many NATO countries at higher levels. In some countries double-digit annual R&D budget increases have been allocated.

Governments in most countries are following a double-track strategy. On the one hand, conventional arms control negotiations are being undertaken with greater seriousness than previously, and negotiated cuts in manpower and equipment are official policy. On the other hand, the process of developing new and more sophisticated weapons and platforms has not been halted. Few major projects have been cancelled, although smaller and lower-priority programmes have been deferred and in several projects the number of systems to be acquired has been reduced.

Without the completion of the fundamental revision of strategy and military doctrine in the context of the NATO alliance, governments continue to invest in weapons development to keep as many future options open as possible. However, growing R&D expenditure cannot compensate companies for major cuts in production. From a company perspective, this might even exacerbate problems in the future if, when today's R&D projects are ready to enter production, either the financial situation or improved international relations in Europe will not allow manufacture to begin.

[7] Legge, M., 'NATO defence planning after CFE', *NATO's Sixteen Nations*, vol. 35, no. 3 (June 1990), pp. 31–35. Legge is NATO Assistant Secretary General for Defence Planning and Policy.

III. Company responses

Companies with an interest in arms production are reacting to the new situation with a variety of different strategies.[8] The most important determinant of company policy is dependence on arms production. Two basic approaches can be distinguished. Companies either try to stabilize, and in some cases expand, their market share, or try to withdraw, partly or totally, from arms production.

Stabilization strategies

1. National and international mergers, take-overs and the formation of new companies

Mergers on a national level and expansion through the acquisition of other companies within the arms industry are not new phenomena and continue to take place, especially in the European aircraft industry, which has merged from dozens of companies to basically one or two major companies in each of the large West European countries. For other reasons a similar concentration process—while at the same time drastically reducing production capacities—has taken place in the shipyards. A new dimension of the concentration process was, however, reached with the Daimler Benz purchase of MBB in 1989 and the formation of the new company Deutsche Aerospace. This merger of the two top arms producers in the FRG made Daimler Benz the third largest arms producer in Europe, approximately of the same size in arms sales as Thomson S. A. in France (see chapter 10). In the UK, GEC bought parts of Ferranti; as a result GEC is the only company able to produce airborne radar.

Cross-border mergers and international take-overs in the defence sector are a somewhat new phenomenon. In terms of mergers, the largest producers—including GEC and Rolls Royce (UK), Thomson-CSF, Matra and Aérospatiale (France), MBB and MTU, both subsidiaries of Deutsche Aerospace (FRG)—have sought to form new companies. In terms of take-overs, the two most important have been the purchase of the British firm Plessey by GEC (UK) and Siemens (FRG), and the take-over of Hollandse Signaalapparaten, a subsidiary of the Dutch firm Philips, by Thomson-CSF (France). While there has been a strong trend of cross-border mergers and take-overs in the civil sector in the past, it is a recent phenomenon in the military sector. For the most prominent examples of this process, see appendix B, tables B.1 and B.2.

In a number of cases in Europe, such co-operation resulted in the dominance of certain market sectors by individual companies. In Europe the main producers of engines, military electronics, missiles and helicopters are in the

[8] Taylor, T., 'The future of European defence industries: problems and responses', paper presented at the British International Studies Association Annual Conference, London, Dec. 1989; Huffschmid, J. and Voß, W., *Militärische Beschaffungen–Waffenhandel–Rüstungskonversion in der EG. Ansätze koordinierter Steuerung*, PIW-Studien no. 7 (Progress-Institut für Wirtschaftsforschung: Bremen, 1991).

process of forming joint companies, and the small arms producers are integrated under the umbrella of GIAT (France) and Royal Ordnance (UK).

2. Strategic alliances and industrial cartels

Not all international co-operation leads to the formation of new companies or take-overs. So-called 'strategic alliances' between companies may increase the probability of securing military contracts in an increasingly uncertain market without sacrificing company autonomy. Major actors who would once have bid alone instead submit joint bids for major contracts or in specific sectors, partly to improve their chances of penetrating markets that are still primarily politically determined and partly to reduce the costs of developing new weapons.

This form of company behaviour has a longer history than mergers between arms producing companies—indeed, in major programmes collaboration has become a way of life. In the case of British Aerospace (the largest arms manufacturing company in Europe), five of its last six military aircraft have been the result of collaboration. These have been produced with both European partners (the Jaguar, the Tornado and the European Fighter Aircraft, EFA) and with US partners (the AV-8B Harrier and the T-45 Goshawk). Only the Hawk jet trainer can be called a single-company endeavour. Recently a new co-operative agreement was signed by Royal Ordnance, a subsidiary of British Aerospace, and the Land Systems Division of General Dynamics. This agreement brings together two of the largest arms producing companies to 'promote sales or lay the foundation for other collaborative projects'.[9]

Governments find it difficult even to discuss multinational decision making. However, the process of moving arms production out of government ownership into the private sector may make the discussion moot, given that the existing economic climate for private sector companies is forcing an internationalization. Perhaps a more interesting question is whether internationalization will automatically lead to industrial rationalization.

The establishment of production consortia does not offset the existence of near-monopolies in the development of major systems in Europe, although larger industrial groupings may be better equipped to challenge US companies in global markets. In the European arms market, the production of helicopters offers the best example of the impact of international production agreements on the rationalization of industry. The central problems of the European helicopter manufacturers were candidly laid out in 1986 by the Chairman of the British firm Westland as 'overmanning, overproduction and lack of profitability'.[10]

The European helicopter industry has undergone an intensive restructuring over the period since 1978, when the defence ministers of France, the FRG,

[9] *Financial Times*, 17 May 1990.

[10] Testimony of Sir John Cuckney before the House of Commons Defence Committee in *The Defence Implications of the Future of Westland Plc*, Report of the House of Commons Defence Committee, 1985–86 (Her Majesty's Stationery Office: London, 1986).

Italy and the UK signed a Memorandum of Understanding committing them to 'work together to develop and produce new helicopters, including their engines and equipment'.[11] The aim was to decide which companies would take the lead in efforts to meet European armed forces' requirements for light, medium and heavy helicopters in land- and shipborne versions. A settlement was sought which would provide sufficient work shares for national industries to remain viable while leaving governments with a degree of choice in procurement.

In an Anglo-Italian consortium, Westland and Agusta were to develop a heavy-lift helicopter through a joint-venture company, EH Industries. Known as the EH-101, this helicopter was in 1991 in the final stage of development. A four-nation consortium involving France (Aérospatiale), the FRG (MBB), Italy (Agusta) and the UK (Westland) was to develop a medium helicopter (the NH-90) under a 1985 Memorandum of Understanding.[12] No agreement was reached on a common light attack helicopter. Agusta and Westland, with limited support from Fokker of the Netherlands and CASA of Spain, attempted to develop an improved version of the Italian A-129 Mangusta for this requirement. This helicopter was to have been called the A-129 Tonal. At the same time, Aérospatiale and MBB have developed the PAH-2 Tigre.

The Netherlands and the UK withdrew from the A-129 programme in November and December, respectively, of 1990, reducing the potential sales of the aircraft by up to 175 helicopters. The requirement will now be met either through the purchase of the US AH-64 Apache or the deferment of orders to the end of the 1990s, when the Franco-German Tigre should be available (the Tigre will not enter production until 1997 at the earliest). In 1990 Aérospatiale (60 per cent share ownership) and MBB (40 per cent share ownership) created a new joint-venture company, Eurocopter, and extended invitations to Agusta and Westland to join their consortium in an effort to improve their chances of winning the contract for future attack helicopters.

This 'top down' effort at rationalization began to break down in 1987 following an internal disagreement within the British Government over whether participation in the European NH-90 consortium should be favoured on political grounds over the cheaper and less risky option of co-producing the UH-60 Blackhawk—already developed by the US firm United Technologies.[13] The UK formally withdrew from the NH-90 project in April 1987.

[11] Declaration of principles to be adopted for co-operation on helicopter programmes by the governments of the French Republic, the Federal Republic of Germany, the Italian Republic and the United Kingdom, reproduced as appendix 1 in *The Defence Implications of the Future of Westland Plc* (note 10).

[12] The Netherlands also has a limited role in the development of the NH-90. Note by the National Armaments Directors of France, FR Germany, Italy and the United Kingdom, 29 Nov. 1985, reproduced as Appendix 6 in *The Defence Implications of the Future of Westland Plc* (note 10).

[13] Secretary of State for Defence Michael Heseltine resigned when Prime Minister Thatcher decided not to oppose the purchase of 26.6 per cent of Westland shares by the US company Sikorsky (a subsidiary of United Technologies). Heseltine believed that he had been authorized to support the formation of a European consortium.

In July 1991 the Italian company Agusta joined the NH Industries consortium which is expected to produce the NH-90. The Italian Government will now provide 27 per cent of the funding for the development of the helicopter. However, the future of the NH-90 was put in doubt in August 1991 when several press reports suggested that France—committed to supply 42 per cent of the funds for the programme—was preparing to suspend for two years allocations to Aérospatiale for the programme.[14] French Prime Minister Edith Cresson quickly denied that the programme was in jeopardy.[15]

The 'Westland affair' could be interpreted as an effort by the UK to reduce the extent of protection of national 'champions' by procurement authorities within the European armaments market and to replace it with a more open competition for government contracts. There may be some truth in this, although, as one recent study has noted, 'the extent of internationalisation can be over-emphasized since British firms continue to take the great bulk of MoD equipment expenditure. Only 10 per cent of the equipment budget goes direct to overseas suppliers although it is quite possible that a higher, and rising, proportion of sub-contracts are being awarded to foreign firms.'[16] The real problem is that the cost of developing and producing helicopters makes large-scale purchases by governments impossible within current procurement budgets. The procurement of helicopters may increase given the current priority attached by European governments to the creation of mobile forces. However, in the near term there is no evidence that governments are prepared to accept the expenditure implications of their new force planning.

3. Producing weapons not affected by arms control and budget cuts

Not all programmes are affected by arms control and budget cuts. Those items not limited by the 1990 Conventional Armed Forces in Europe (CFE) Treaty, primarily missiles, might actually experience a growth in sales, which would compensate for cuts in other areas. Reductions in traditional areas of the military budget, especially investments in weapon platforms, may be compensated for by high technology. Producers of military electronics, telecommunications and data-processing equipment—often key technologies in modernization programmes—are likely to be less affected than companies making the 'platforms', that is, the weapon carrier systems.

4. Increased exports

While the export of major conventional weapons has been shrinking, some European companies nevertheless hope to expand their export sales—partly at

[14] *Defense News*, 26 Aug. 1991, p. 3.
[15] *Atlantic News*, 29 Aug. 1991, p. 4.
[16] Southwood, P., *The UK Defence Industry at the Crossroads: An Analysis of the Industrial Impact of Options for Change Study and the CFE Process* (Enterprise for Defence and Disarmament: Chippenham, Wiltshire, 1990), p. 36.

the expense of other companies, and partly in those systems whose sales are still experiencing growth. Especially British and French companies (British Aerospace, Ferranti, Thomson and DCN), but also Blohm & Voss in the FRG, have announced such plans or are involved in large export programmes.

Another growing tendency in foreign sales is the retro-fitting or upgrading of weapon and ancillary systems on existing platforms, avoiding the need to invest in the high-cost development of new major platforms. As the unit costs of new systems continue to increase and procurement budgets fall in real terms, this has become an increasingly attractive option for many governments.

Adjustment strategies

1. Lay-offs and shrinkage

Several of the strategies mentioned above often involve reductions in employment. While the practice of 'hire and fire' is more common in the USA, where tens of thousands of employees in several companies have been given notice recently,[17] it is not unusual in Western Europe. Companies have already fired employees or announced reductions to take place during the next few years.

This is true for companies in many countries, most prominently in France by Dassault, GIAT and Aérospatiale; in the UK by British Aerospace, GEC, Westland, Short Brothers, Vickers, VSEL and Vosper Thornycroft; in the FRG by MBB of Daimler Benz and Rheinmetall; but also in Sweden by Bofors and in Belgium by FN and PRB. In several cases, however, it was possible to move the affected workers to non-military production within the companies.[18]

2. Sale of arms production facilities

In addition to lay-offs, some companies are trying to sell their arms production facilities, or subsidiaries, or to consider plant closures. The most prominent example of a sale of arms producing plants was the sale of the Philips subsidiary Hollandse Signaalapparaten to the French company Thomson-CSF. Another example is the sale of certain divisions of the British company Ferranti-International Signal after the US subsidiary of the company was charged with fraud. Presently the management of Oerlikon Bührle of Switzerland has decided to sell off its arms production subsidiaries. British Aerospace has already closed several plants. With reduced business prospects, it is not always possible to sell arms production units. The British electronics company Thorn EMI, whose annual arms sales of over $500 million in 1989 have brought the company into the group of the largest European arms producers, could not sell its arms producing division in 1989.

[17] Wulf, H., 'Arms production', SIPRI, *SIPRI Yearbook 1991: World Armaments and Disarmament* (Oxford University Press: Oxford, 1991), p. 290.

[18] For an estimate of the jobs lost in the West European arms industry see below.

3. Diversification or conversion into civil production

Arms producing firms are responding to the new market situation with two types of diversification strategy. The first is to reduce dependence on arms contracts by acquiring civilian-oriented companies. Two typical recent examples are the acquisition by British Aerospace of the automobile producer Rover, and the acquisition by the West German artillery producer Rheinmetall of the automobile spare parts producer Pierburg.

The second strategy is to convert production capacities from military to civilian production. A number of companies (i.e., most shipyards as well as tank producers such as the German firms Diehl and Krauss-Maffei) have already begun to reorient their capacities as a result of decreased orders for particular weapon systems. Often companies try to make use of skills and technology acquired in the arms production process for their non-military activities. Naturally, there is no general rule for the effectiveness of such a process, and the record shows that both failures and successes have occurred. The more specialized a company is, the more difficult it is to move into other areas, especially into civil markets that are already highly competitive.

4. Disarmament-related diversification

Also of interest are two areas that emerge as a result of arms control and disarmament: verification and the destruction of arms. Producers of electronics, optronics and other specialized equipment will benefit from the need for a variety of verification technologies. It is too early to estimate what the arms control process will have to offer these companies since the size of the procurement will depend on how far the arms control process goes and to what extent and precision verification is required. The need to verify dismantling or destruction of weapon systems to prove compliance with arms control agreements requires new investment. Both the early experiences with the INF Treaty and the destruction of chemical weapons suggest that it will be more expensive than originally anticipated.[19] Disarmament and the verification of arms control agreements could compensate a number of high-technology companies for lost business in arms production. Some companies, such as Diehl, are offering their services to destroy surplus weapon systems. Smaller companies have specialized in dismantling surplus munitions. The destruction and conversion of weapons are of only marginal benefit to the macro-economy and only in rare cases are modern weapon systems of use for non-military purposes. However, for a few companies the disarmament process offers new business opportunities, since tens of thousands of superfluous major weapon systems will have to be destroyed throughout Europe, mainly in the USSR.

[19] See Griffiths, S. I., 'The implementation of the INF Treaty', and Lundin, S. J. and Stock, T. 'Chemical and biological warfare: developments in 1990', SIPRI, *SIPRI Yearbook 1990: World Armaments and Disarmament* (Oxford University Press: Oxford, 1990), pp. 443–58 and 85–112, respectively.

While reducing military procurement might be beneficial to the economy, especially compared to the decades of armament competition, the dismantling or conversion of major weapon systems actually represents the destruction of products that have been manufactured at great cost to the economy.

IV. The impact of the CFE Treaty on the West European arms industry

The direct impact of the CFE Treaty on the West European industrial base will be negligible. The limited new markets which will be created by disarmament-related activities were noted above. The implementation of the CFE Treaty is also unlikely to have significant direct result on those companies manufacturing treaty-limited items.

The limited impact of the CFE Treaty is not surprising given that negotiating positions advanced by NATO took account of existing plans for force modernization in the alliance. While requiring major reductions in the level of Soviet armed forces, the CFE agreement will have very little direct impact on currently planned modernization programmes within NATO. Moreover, in any follow-on process, NATO defence ministries will again try and ensure that arms control does not prevent the development, production and procurement of systems considered necessary for the effective operation of the armed forces.

Although discussed at earlier stages of the CFE Negotiation as an element of the verification process, controls over or the monitoring of arms production are not mentioned in any of the articles dealing with verification (articles XIII, XIV and XV). Arms produced for export are exempt from Treaty counting rules under Article III, paragraph 1E, which states that items that 'are awaiting, or are being refurbished for, export or re-export and are temporarily retained within the area of application' are not considered as treaty-limited equipment.[20]

A small number of European companies are likely to be affected by the ceilings applied within the CFE Treaty to holdings of specified treaty-limited items. However, the companies which may be affected include some that are of disproportionate importance to certain industrial sectors and to technology development within their respective countries.

The indirect impact of the CFE Treaty is likely to be more important. In particular, the transfer of arms that cannot be accommodated within the country sub-limits on treaty-limited items set for NATO countries within the CFE Treaty may reduce demand for new major conventional weapons. These items will be transferred to other NATO allies under what has become known as the NATO Equipment Transfer and Equipment Rationalization Programme but which was formerly called the 'cascade'.[21]

[20] The Treaty on Conventional Armed Forces in Europe (the CFE Treaty) is reproduced in full in SIPRI (note 17), appendix 13A, pp. 461–74. For Article III, paragraph 1E, see p. 464.

[21] The cascade is described in SIPRI (note 19), p. 318, and in SIPRI (note 17), pp. 426–27.

Table 2.4. Equipment transfer programme anticipated under the NATO 'cascade'

Receiver country	Treaty-limited item	No. to be transferred[a]	Supplier country
Denmark	Leopard 1 tank	110	FRG
	M-110 artillery	36	USA
FRG	Leopard 1 tanks	400	Denmark, Greece, Turkey
	M-113 ACV	550	Greece, Turkey
	110-mm rocket launchers	150	Greece, Portugal, Turkey
Greece	Various tanks	700	USA, FRG, Netherlands
	Various ACVs[b]	150	USA
	M-110 artillery	70	USA
Italy	M-113 ACV	100	Turkey
Netherlands	Leopard 1 tank	170	Greece
	M-113 ACV	130	Portugal
Norway	Leopard 1 tank	90	. .
	M-113 ACV	125	USA
	M-113 ACV	100	Netherlands
Spain	M-60 series tanks	530	USA
	M-113 ACV	100	USA
Turkey	Various tanks	1 050	USA, FRG
	Various ACVs	600	USA, FRG
	M-110 artillery	70	USA
USA	M-60 series tanks	2 000	Greece, Portugal, Spain, Turkey
	M-113 ACV	600	Greece, Portugal, Spain, Turkey
	M-110 artillery	180	Denmark, Greece, Turkey

[a] All numbers are approximate.

[b] Armoured combat vehicles.

Source: *Jane's Defence Weekly*, 6 July 1991, p. 18.

As indicated in table 2.4, some of the countries which are to receive significant quantities of surplus equipment under the NATO 'cascade' are those which were previously seen as important export markets for both European and US companies. Portugal and Turkey in particular are two of the few countries whose expenditure on major equipment has increased in recent years.

Finally, once implemented and verified, the CFE Treaty will contribute to a more benign threat environment that may influence future decisions about the size, shape and equipment fit appropriate for European armed forces.

The industrial impact

On the assumption that broadly the same group of companies which supplied items now subject to CFE cuts would also supply the next generation of the same items, the number of major companies which could suffer lost orders as a direct consequence of the CFE Treaty probably numbers no more than 50.

Table 2.5. European NATO military equipment holdings in November 1990, cuts required and 1995 CFE ceilings

Country		Battle tanks	Artillery	ACVs	Combat aircraft	Attack helicopters
Belgium	Holdings	359	376	1 282	191	–
	Cuts required	25	56	183	–	–
	CFE ceiling	334	320	1 099	232	46
Denmark	Holdings	419	553	316	106	3
	Cuts required	66	–	–	–	–
	CFE ceiling	353	553	316	106	12
France	Holdings	1 358	1 330	4 125	700	429
	Cuts required	52	38	305	–	77
	CFE ceiling	1 306	1 292	3 820	800	352
FRG	Holdings	7 133	4 644	9 598	1 064	357
	Cuts required	2 967	1 939	6 152	164	51
	CFE ceiling	4 166	2 705	3 446	900	306
Greece	Holdings	1 725	1 941	1 639	480	–
	Cuts required	–	63	–	–	–
	CFE ceiling	1 735	1 878	2 534	650	18
Italy	Holdings	1 912	2 222	3 591	584	169
	Cuts required	564	267	252	–	27
	CFE ceiling	1 348	1 955	3 339	650	142
Netherlands	Holdings	913	838	1 467	196	91
	Cuts required	170	231	387	–	22
	CFE ceiling	743	607	1 080	230	69
Norway	Holdings	205	532	146	90	–
	Cuts required	35	5	–	–	–
	CFE ceiling	170	527	225	100	–
Portugal	Holdings	146	334	259	96	–
	Cuts required	–	–	–	–	–
	CFE ceiling	300	450	430	160	26
Spain	Holdings	854	1 373	1 259	252	28
	Cuts required	60	63	–	–	–
	CFE ceiling	794	1 310	1 588	310	71
Turkey	Holdings	2 888	3 202	1 554	589	–
	Cuts required	93	–	–	–	–
	CFE ceiling	2 795	3 523	3 120	750	43
UK	Holdings	1 198	636	3 193	842	368
	Cuts required	183	–	17	–	–
	CFE ceiling	1 015	636	3 176	900	384
Total NATO Europe	Holdings	19 110	17 981	28 429	5 190	1 445
	Cuts required	4 051	2 225	4 256	–	–
	CFE ceiling	15 059	15 756	24 173	5 788	1 469

Source: Sharp, J. M. O., 'Conventional arms control in Europe', SIPRI, *SIPRI Yearbook 1991: World Armaments and Disarmament* (Oxford University Press: Oxford, 1991), p. 425.

These companies are concentrated in the heavy engineering sector, since the largest reductions will have to be made in European holdings of main battle tanks, artillery and armoured vehicle holdings. Table 2.5 shows the holdings of European NATO countries in the categories of items limited by the CFE Treaty, together with the stipulated ceilings and the numbers to be cut. It is apparent from the table that the country which is making by far the greatest cuts in forces is the FRG, followed by Italy and the Netherlands.

Given that the market for military equipment is still dominated by national producers, arms producing companies in the heavy engineering sector in these three countries may therefore be most affected by the CFE Treaty cuts. Companies such as Diehl, Krauss-Maffei, MAK and Oto Melara, all heavily dependent on defence contracts, may face bleak future prospects.

If companies producing tanks, heavy armoured vehicles and heavy artillery pieces are generally likely to suffer the most direct market losses, some of these companies may gain from the fact that the mobile forces which France, Italy, Spain and the UK are now planning to raise will need new equipment.

Mobile forces require as much firepower as can be carried without loss of speed of movement. For example, the British firms Royal Ordnance (in collaboration with the US firm Bowen McLaughlin York) and VSEL are developing 155-mm howitzers that can be carried by medium transport helicopters such as the UH-60 Blackhawk. These lightweight howitzers use new composite materials and plastics to replace heavy metal parts. The prototypes were built as private ventures, indicating that company thinking about future developments in force structure is considerably in advance of many academics and politicians.

Nevertheless, the creation of these mobile forces is dependent on the progress of equipment programmes intended to provide the volume of lift capacity which will allow them to move—whether by sea or by air. Assault ships and especially large transport aircraft are two areas where European armed forces remain heavily dependent on the USA. At the level of tactical transport, the creation of mobile formations depends on the procurement of significant numbers of medium- and heavy-lift helicopters.

V. The future of the industry: a medium-term prognosis

In order to provide an estimate of the order of magnitude of the possible future economic situation facing the arms industry, two scenarios are presented below. From each scenario, two projections are made—one estimating procurement expenditure and one estimating employment. The two projections rest on the following assumptions:

In scenario 1 it is assumed that: (*a*) annual reductions in procurement spending will amount to 3 per cent (this assumption extrapolates the trend over 1985–90); (*b*) the CFE Treaty will not have a great impact on present programmes; (*c*) the volume of the trade balance (export minus import) of

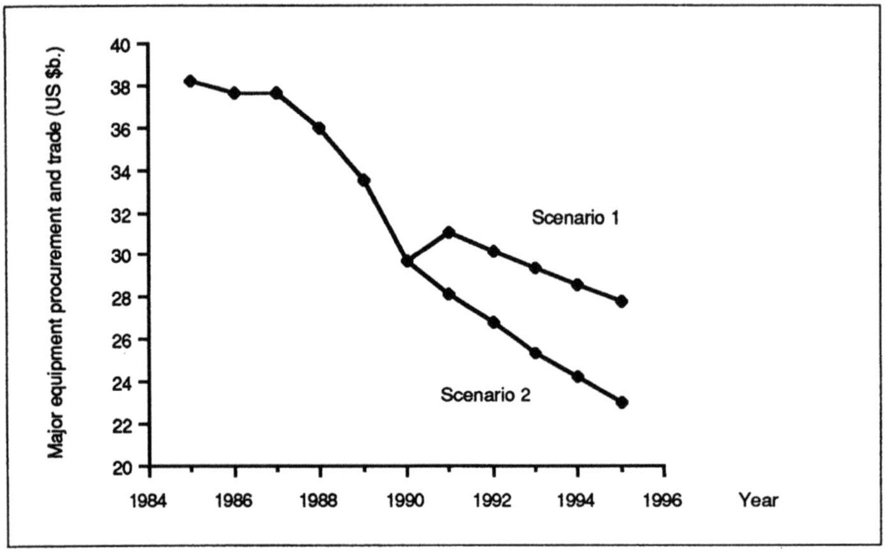

Figure 2.3. Actual and potential domestic procurement of major equipment and major arms exports of European NATO countries, 1985–95

Source: SIPRI data base.

major arms of NATO Europe will stabilize at the level of 1985–89 as recorded by SIPRI;[22] and (*d*) annual productivity gains in the arms industry will amount to 2 per cent.

In scenario 2 it is assumed that: (*a*) a second CFE treaty will require deep cuts in conventional arms in NATO countries; (*b*) reductions in procurement budgets will accelerate to 5 per cent annually as a result of the international climate and successful arms control; (*c*) competition in the world arms market will increase and the European NATO countries' trade balance in major conventional arms will shrink by 5 per cent annually; and (*d*) as in scenario 1, annual productivity gains in the arms industry will amount to 2 per cent.[23]

[22] Exports minus imports of major conventional weapons in the European NATO countries amounted to an annual average of $2 billion for the period. In 1990 this figure was 0 due to fluctuations in sales. This assumption results in an increase of exports for 1991, reflected in an upswing of the curve in scenario 1 for the year 1991.

[23] A note of caution is required on the applied methodology and thus the precision of this estimate. First, the procurement figures and the arms export figures do not include the total of arms production and exports but only of major equipment. In the calculation it is therefore implicitly assumed that the remaining part of arms production and exports follows the same pattern as for major equipment. Second, the arms procurement and trade figures are not really comparable for several reasons: procurement figures are given at 1988 constant prices, and arms trade at 1985 constant prices. The SIPRI arms trade statistics are trend indicators of the deliveries of major conventional weapons and not figures which measure what was actually paid for arms supplied. Nevertheless, these trade figures are a realistic reflection of the export trade in major arms on the part of European NATO countries. Finally, the employment figures of the mid-1980s were not based on detailed input–output studies—since such data were not available—but are estimates based on several sources. In conclusion: the figures in both scenarios are not a precise prognosis of future developments but estimates of the possible economic dimensions of arms control and disarmament in the European NATO countries as they affect the arms industry.

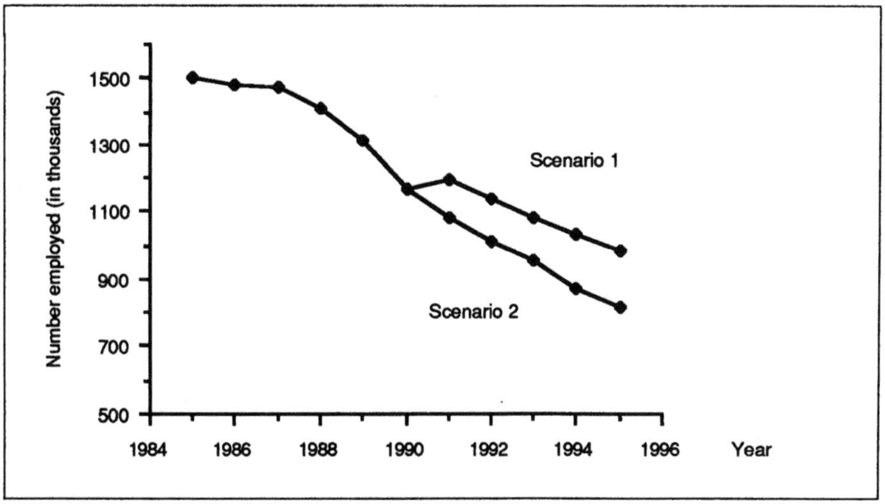

Figure 2.4. Actual and potential employment in the arms industry of European NATO countries, 1985–95

Source: SIPRI data base.

Scenario 1 is a conservative estimate that discounts major revisions in current programmes, while scenario 2 is perhaps more realistic as it rests on the premise that a fundamental change in procurement policy is needed.

The result of the projection is presented in figures 2.3 and 2.4. The value of procurement of major items of equipment plus the trade balance of major conventional weapons would fall to $27.8 billion by 1995 in comparison with $33.5 billion in 1989. The total value of major equipment produced in European NATO countries will go down by about one-sixth by 1995. Under scenario 2 the value of procurement of major items of equipment would fall to $23.0 billion by 1995—a reduction of almost one-third. In reality, these projections probably represent a range within which future expenditure is likely to fall, since cuts envisaged in scenario 1 can be considered the minimum probable while reductions are unlikely to exceed those predicted in scenario 2.

The impact on those West Europeans whose employment is directly dependent on arms production will be even greater. A total of 150 000 jobs were lost in arms production in 1988–89, and another 150 000 in 1990. Additional job losses in the range of 185 000–350 000 can be expected in the period 1991–95. From the mid-1980s level of 1.5 million, the number of jobs directly dependent on arms production in the European NATO countries will by 1995 fall to 980 000 in scenario 1 and to 815 000 in scenario 2.[24] In scenario 2, employment in the arms industry in European NATO countries would almost be halved in the period 1985–95.

[24] Since the reduction of arms production in Europe was larger than was anticipated in the first estimate (see note 1), the job losses reported here are larger than those predicted in 1990.

3. Institutional frameworks for integration of arms production in Western Europe

Harald Bauer

I. Introduction

Since the early 1970s numerous attempts have been made to co-ordinate West European arms production. None of these efforts have met with any real success, reflecting the disparity of the goals and interests of the various governments involved. While US unwillingness to accept an integration of the West European arms market has been the main obstacle, the uneven distribution of production capacities and market shares among West European countries has given rise to friction, in the form of competitive national policies, within Western Europe itself. The British and French arms industries have long held a lead, a position the two governments have tried to safeguard while other West European countries have sought to improve theirs.[1]

To improve the cost-effectiveness of arms co-production, different institutional frameworks were tried. The Western European Union's (WEU) Standing Armament Committee, NATO's Conventional Armaments Planning System (CAPS) and Conference of National Armaments Directors (CNAD) and the Independent European Program Group (IEPG, originally set up close to NATO) initiated many working groups, whose studies and proposals filled thousands of pages. Multi- and bilateral approaches were tried, but neither cost control proposals nor material results were convincing to the governments involved.

This chapter concentrates on two specific actors, the European Community (EC) and the IEPG. The other bodies mentioned above have been weakened by developments in the 1980s. Much of the WEU's capacity is paralysed by a long-lasting internal conflict over its role, competence and the location of its secretariat, leaving, among other things, its Standing Armament Committee short-staffed.[2] NATO's CNAD has not promoted new projects since the fall of

[1] On the problems of traditional arms co-production, see Wulf, H., 'Europäische Zusammenarbeit in der Rüstungsproduktion: Ziele, Probleme und Perspektiven', eds L. Brock and M. Jopp, *Sicherheitspolitische Zusammenarbeit und Kooperation der Rüstungswirtschaft in Westeuropa* (Nomos: Baden-Baden, 1986), pp. 107–28; *Report drawn up on behalf of the Political Affairs Committee on European Armaments on Procurement Cooperation*, E. Klepsch rapporteur ('Klepsch Report'), document no. 83/78 (Office for Official Publications of the European Communities: Luxembourg, 8 May 1978).

[2] 'European co-operation in armaments research and development: guide-lines drawn from the colloquy', in *Report submitted on behalf of the Committee on Scientific, Technological and Aerospace Questions*, Mr Wilkinson, Chairman and rapporteur, Assembly of the Western European Union, document no. 1141 (Western European Union: Paris, May 1988), p. 15; 'The reactivation and enlargement of WEU: reply to the annual report of the Council', submitted on behalf of the General Affairs Committee by Mr

the Berlin wall in November 1989; the National Armament Directors are now acting primarily within the IEPG.

There are several reasons for the shift away from NATO forums. Following France's lead, reluctance to accept continued US domination of the West European arms market grew among the European allies during the 1980s. Since the beginning of the decade, when their combined economies matched that of the USA, two different debates have surfaced. One had a political and military focus, expressed by the challenge that the European pillar of NATO needed to be strengthened. The other had an economic focus, centred on a perceived 'Eurosclerosis', the widespread feeling that the industries and governments of Western Europe were half asleep and that they risked losing their competitiveness in key high-technology fields. This would also have serious implications for the independence of the European arms industries. It seemed only natural, therefore, that the countries of Western Europe sought to find new ways to safeguard what they felt to be common national interests.

At a first glance, the EC seems the logical forum for such co-operation. The EC is creating a Single Market, a coherent economic space among its member countries, including common regulations, standards and company law. There is no inherent reason why arms procurement and production should be exempt from the regulations and deregulations of the Single Market, other than the fact that states regard the control over arms production and procurement as issues of sovereignty. For the USA the impact of the Single Market on arms trade and co-operation is a matter of concern, as reports to Congress and other studies prove.[3] Furthermore, the 1990 discussion about and proposals for a European political union have added momentum to the EC's intention to have all aspects of arms production under its control. It is simply not conceivable that a political union of the 12 EC member states would preclude the armaments sector.

In the autumn of 1988, after several years of preliminary work, the IEPG adopted a so-called Action Plan to create a (Western) Europe-wide arms market (see appendix F).[4] However, the IEPG has little organizational back-up and no means to impose common regulations on national authorities. Until 1990 the ambassadors to NATO co-ordinated the activities of the IEPG.

II. The European Community

The provisions of the Treaty of Rome concerning internal market regulations do not apply to the arms industries of the EC member states. Article 223 of the 'General and Final Provisions' of the Treaty establishes that:

van den Sanden, rapporteur, Assembly of the Western European Union, document no. 1163 (Western European Union: Paris, Nov. 1988), p. 10.

[3] Edwards, M. B., *EC 1992: Potential Implications for Arms Trade and Cooperation*, Congressional Research Service Report no. 89-642F (US Library of Congress: Washington, DC, 3 Nov. 1989); Office of Technology Assessment, *Arming our Allies: Cooperation and Competition in Defense Technology* (US Government Printing Office: Washington, DC, May 1990), p. 55.

[4] Independent European Programme Group, *Action Plan on a Stepwise Development of a European Armaments Market*, document no. NAD/D-22 (IEPG: Brussels, 23 Sep. 1988).

(a) No member state shall be obliged to supply information the disclosure of which it considers contrary to the essential interests of its security;

(b) Any member state may take such measures as it considers necessary for the protection of the essential interests of its security which are connected with the production of or trade in arms, munitions and war material; such measures shall not adversely affect the conditions of competition in the common market regarding products which are not intended for specifically military purposes.[5]

To this end, the European Economic Community Council in 1958 adjoined to Article 223 an unpublished list of products to which these provisions apply (see appendix E). The EC Commission has the right to propose changes to this list, but for them to be adopted the Council must agree unanimously on all changes. The list has not been changed since 1958.

The 1986 Single European Act implements no new regulations limiting the validity of Article 223 of the Treaty of Rome, but in Title II and III of the Act there are some articles that can be interpreted as aiming directly or indirectly at a greater liberalization and harmonization of arms procurement. Article 130, paragraphs f–q, provide the EC with an enhanced role for the implementation of research and development (R&D) programmes, in theory also applicable to arms procurement, under the terms of industrial policy. Article 30.6 of Title III states that the contracting parties consider that closer co-operation on questions of European security would contribute in an essential way to the development of a European identity in external policy matters. The same Article confirms the right of member states to maintain the technological and industrial conditions necessary for their security, with the provision that efforts to that end, where appropriate, may also be made within the framework of the competent institutions and bodies. These are not specified; what is meant are non-EC bodies, such as the WEU and NATO.

This approach of claiming competence in the field of arms production and procurement in the name of industrial policy has been proposed on several occasions by different actors within the EC. The 1978 Klepsch Report to the European Parliament[6] suggested that the EC follow that route, as did the 1980 Greenwood Report to the EC Commission.[7] The same approach was recommended by two consecutive EC commissioners responsible for industrial affairs and the EC internal market. In a lecture given in 1987 at the General Staff Academy of the German Bundeswehr, Commission Vice-President Karl-Heinz Narjes expressed at length the importance of strengthening the technological and industrial base of Western Europe in order for it to remain competitive *vis-à-vis* the USA and Japan. He stressed in particular the importance of EC policy to maintain and promote, through its research programmes, a broad scale of

[5] The article is reproduced in full in appendix E. See also 'European armament cooperation after the implementation of the Single European Act', unpublished mimeograph, European Parliament, Directorate General for Research, Political and Institutional Affairs Division, Luxembourg, 1988.

[6] Note 1.

[7] 'Report on a policy to strengthen the defence and technological cooperation among the Western European countries', prepared by David Greenwood for the European Commission, document no. CAB/XII/175/80/DE (Commission of the European Communities: Brussels, 1980).

high-technology skills. In his view, this is a precondition for safeguarding an independent arms industrial base on a European scale.[8] In March 1989 the new Commissioner, Martin Bangemann, restated this view before the European Parliament, adding that the general contractual directions would also apply to arms production and trade.[9] In the view of the Commissioner's Directorate General for the Internal Market, it is becoming difficult to justify that two independent frameworks are needed to maintain civil and defence-oriented industries. For economic as well as technological reasons, these spheres of industrial activity are growing increasingly interdependent.

The Commission has presented two other arguments along the same line. Regarding so-called dual-use technology, it argues that in many areas (in particular electronics) it is no longer possible to distinguish between civil and military items, and that in such instances this distinction cannot be used to determine the Commission's area of competence and, by extension, the applicability of EC regulations. A more recently advanced argument holds that any benefit of a 'peace dividend', should there indeed be one, may not be used by national governments for the regional conversion of arms industries, as regional policy is within the Commission's area of competence.[10]

In practice, the Commission's efforts have not yet been very successful. Its amendment to Directive 77/62 on the co-ordination of procedures for awarding public supply contracts was intended to cover arms contracts as well, with the exception of articles listed in the appendix to Article 223.2. France, Italy and Greece objected, however, and the arms sector was not included. A similar fate met the Commission's suggestion of a Council regulation to suspend import duties on certain weapons and military equipment. The Commission's intention was to harmonize, under its authority, the arms import regulations of member states and to establish a common external tariff on military equipment not listed in the appendix to Article 223.2.[11] A commission of the EC Council now is charged with finding a solution to this problem. The 1992 Single Market requires the introduction of common or at least co-ordinated regulations on arms exports because all other goods will be covered by that market and controls at inner EC borders will be dropped.[12]

The procurement of 'soft' military equipment not excluded from EC competence by Article 223.2 is subject to EC regulations on public procurement. According to Commissioner Bangemann, in 1989 the *Official Journal of*

[8] Narjes, K.-H., 'Die Europäische Einigung am Beispiel der Europäischen Gemeinschaft unter Berücksichtigung ihrer sicherheitspolitischen Aspekte', unpublished mimeograph copy of lecture given at the General Staff Academy of the Bundeswehr, Hamburg, 1987, p. 18.

[9] 'Ausführliche Sitzungsberichte', unpublished mimeograph, European Parliament, Strasbourg, 14 Mar. 1989, p. 28.

[10] 'Ausführliche Sitzungsberichte', unpublished mimeograph, European Parliament, Strasbourg, 10 July 1990, pp. 103 and 172.

[11] 'European armament cooperation after the implementation of the Single European Act' (note 5), p. 14.

[12] See Courades Allebeck, A., 'The European Community and arms export regulations', ed. I. Anthony, SIPRI, *Arms Export Regulations* (Oxford University Press: Oxford, 1991), pp. 212–18.

the European Communities published 1688 calls to tender, of which 1584 concerned supplies to the armies.[13]

Of quite a different scope are the initiatives taken in 1990 to achieve a political union, fixed as the ultimate goal of the EC's progressing integration policy. In its comment on these initiatives, the Commission clearly stated that security and defence matters must be included and that Article 223 be dropped.[14] In their common letter, published before the European Council summit meeting in Rome on 14–15 December 1990, French President François Mitterrand and German Chancellor Helmut Kohl recommended that a political union should also cover foreign and security policy. Foreign Minister Gianni de Michelis of Italy, who held the Council chairmanship at the time, suggested that the EC should take over the organization and tasks of the WEU. Ireland, Denmark and the UK voiced strong opposition to these proposals.[15] The final declaration on the installation of the Intergovernmental Conference on Political Union remains vague on the issue of foreign and security policy.[16] Dramatic changes in this context are not likely in the near future.

III. The Independent European Programme Group

During its first 11 years (1976–86) the IEPG did not stand out from other bodies promoting co-operation in the arms sector. Closely co-ordinated with—but formally not linked to—NATO, the IEPG allowed France, which had left NATO's integrated military structure in 1967, to participate in planning and programmes. The initial purpose of the IEPG was to improve West European arms procurement co-operation, in order to strengthen arms contractors competing with larger and technologically more advanced US firms for contracts. This was primarily a response to US claims for more sophisticated European armaments, a goal to be reached by the standardization of NATO equipment.

In the first half of the 1980s, Dutch and British defence officials tried to invigorate the IEPG. As the British had introduced competition in the arms sector by the so-called Levene reforms (named after former National Armaments Director Sir Peter Levene), they had a strong interest in extending these reforms to Western Europe. Another West European goal was to face new US and Japanese technological challenges, the former explicitly associated with the Reagan Administration's Strategic Defense Initiative (SDI). The Vredeling Report, discussed below, summarizes the arguments:

> For the future, massive US efforts stemming *inter alia* from the SDI initiative could pull that nation rapidly ahead in electronics and other high technologies. In

[13] Institut de Relations Internationales et Stratégiques, *Numéro Spécial* (IRIS: Paris, Oct. 1990), p. 25.

[14] *Commission Opinion of the 21 October 1990 on the Proposal for Amendment of the Treaty Establishing the European Economic Community with a View to Political Union*, document no. COM (90) 600 final (Commission of the European Economic Communities: Brussel, 23 Oct. 1990), p. 5.

[15] Lemaitre, P., 'La réunion de Douze à Venise: vers une politique ètrangère commune . . . à petits pas', *Le Monde*, 9 Oct. 1990.

[16] See *Presse und Informationsamt der Bundesregierung*, Bulletin no. 149 (21 Dec. 1990), p. 1553.

development and production, Europe is less competitive than her current technology base would suggest, primarily because of her fragmented markets. The threat from Japan is not direct, since is does not currently export defence products. There is, however, a significant indirect threat. There are very important transfers of military technology from Japan to the United States. Furthermore, Japan exports large quantities of highly competitive civil high technology products to Europe and this has an undermining effect on the European industrial base which supports both the civil economy and the defence industry.[17]

In 1984 defence ministers replaced NATO ambassadors as senior representatives to the IEPG. Co-operative technology programmes were defined and the first steps taken to enhance research. The European Defence Industry Study, often named after its chairman, former Dutch Defence Minister and EC Commissioner Henk Vredeling, was commissioned in 1985 and drawn up in 1986. Its most important recommendations were: (*a*) to gradually open national arms markets to competition in order to establish a single European arms market; (*b*) to initiate a research programme to strengthen the defence industrial base; (*c*) to drop the strict allocation of the benefits of co-operative projects in shares proportionate to cost shares in favour of the more flexible principle known as *juste retour*, used by the European Space Agency, that aims at mid-term industrial offsets; (*d*) to enhance the transfer of technology within IEPG members; (*e*) to support member countries with developing arms industries (Greece, Portugal and Turkey), and during a period of transition to protect these against the negative effects of competition; and (*f*) to set up a small secretariat to realize these plans.[18] After some delay, partly because of a short period of inactivity following the resignation of British Defence Minister Michael Heseltine,[19] defence ministers approved in November 1988 the *Action Plan on a Stepwise Development of a European Armaments Market*, based on the recommendations of the Vredeling Report and drawn up by the National Armaments Directors (NAD), who make up panel III of the IEPG (Procedures, Economic Matters and Action Plan). The Action Plan, which is reproduced in full in appendix F, outlined the first steps toward a more open market, including the establishment of the 'European Co-operative Long-Term Initiative in Defence' (EUCLID), chaired by France, and a secretariat based in Lisbon. In November 1990 a Memorandum of Understanding regulating EUCLID projects was signed by IEPG Ministers. Around 11 Common European Priority Areas (CEPA) contracts for projects, called Research and Technology Programmes

[17] Independent European Programme Group, *Towards a Stronger Europe*, A Report by an Independent Study Team Established by Defence Ministers of Nations of the Independent European Programme Group to Make Proposals to Improve the Competitiveness of Europe's Defence Equipment Industry ('Vredeling Report') (IEPG: Brussels, 1987), vol. 1, p. 2.

[18] Vredeling Report (note 17), vol. 1, pp. 3–15.

[19] See Creasey, P., 'European defence firms in cooperation agreements', eds P. Creasey and S. May, *The European Armaments Market and Procurement Cooperation* (Macmillan: London, 1988), p. 138.

(RTPs), were arranged on a case-to-case basis, under the auspices of a lead nation.[20]

The opening of the markets of the IEPG member countries seems more important, however. The first steps in this direction include: (*a*) the designation of focal points inside national procurement agencies at which companies from other IEPG countries can register as potential suppliers; (*b*) the harmonization of procurement procedures and criteria; (*c*) the compilation of data by the Permanent Secretariat on national bidding opportunities, potential suppliers and contracts awarded on a cross-border basis; and (*d*) the publication of bulletins on national bidding opportunities on a regular basis. At a meeting of IEPG defence ministers in February 1990 some progress in these areas was recorded, although some countries had yet to meet obligations,[21] and the communiqué of the Copenhagen meeting in November 1990 does not give any precise information on the state of the art. Yet the *Policy Document on the European Defence Equipment Market*, approved in Copenhagen, says that exceptions to the rule of IEPG-wide invitations to tender will be kept to a minimum (see appendix F).

It should be noted that the segment of the arms market opened for IEPG-wide competition is only on the order of 10–20 per cent of national markets for 'hard' defence equipment. In line with the 1987 Anglo-French Reciprocal Purchase arrangement,[22] the regulations apply to production contracts valued at more than £50 million ($82 million) and development contracts valued at over £10 million ($16.4 million). Harmonization of procurement procedures is difficult and wrought with misperceptions. In more than one country there is concern that the government is opening markets ahead of other countries, thus putting domestic companies at a disadvantage.[23]

Nevertheless, the IEPG can be credited with having established the embryo of a West European arms market, something many in the not too distant past had judged impossible. Now the question remains whether political decision makers will leave the further development of this market to an organization as weak and as separate from other institutions as the IEPG. The most important competitor in creating a West European arms market obviously remains the EC.

IV. Rivalry or co-operation?

The outcome of the evident rivalry between the IEPG and the European Community cannot be predicted. The Single Market notwithstanding, as long as Article 223 of the Treaty of Rome remains unchanged, the EC Commission cannot hope to gain full control over the European arms market. Also, defence

[20] See Independent European Programme Group, *Copenhagen Communiqué*, document no. MIN/D-14 (IEPG: Copenhagen, 16 Nov. 1990); *Wehrtechnik*, vol. 22, no. 12 (Dec. 1990), p. 76; *Armées d'aujourd'hui*, no. 155 (Nov. 1990), p. 46.

[21] *Wehrtechnik*, vol. 22, no. 4 (Apr. 1990), p. 15

[22] Stewart, J. M., 'The European defence market—principles and policies', *NATO's Sixteen Nations*, vol. 32, no. 8 (Dec. 1989–Jan. 1990), p. 20.

[23] See answers to questionnaire sent to companies, *Wehrtechnik*, vol. 22, no. 4 (Apr. 1990), pp. 19–25.

ministers deliberately chose the IEPG for the defence market, thus avoiding debates on the civil character of the EC. Another rationale was not to lose control over 'hard' defence items, as 'soft' items are already subject to EC regulations on public procurement. In this view, 'hard' arms are the last procurement stronghold of national defence ministries.

The political obstacles to full implementation of the EC option put forward in the Klepsch Report remain: the question of how to maintain Irish neutrality, and the unwillingness of some governments to introduce military elements into the EC or to give the Commission control over national procurement policies. Only the creation of a political union would bring about fundamental changes. In 1991 such an event seems far from certain.

On the other hand, the IEPG arms market could not exist without the EC framework and the 1992 Single Market. It is the EC that establishes the regulations necessary for the IEPG market, such as common regulations for bidding opportunities, patenting, mergers and acquisitions, technical and legal standardization and European company law. In the words of Mr Battistella, a representative of the French missile producer Matra, 25 per cent of the time necessary to fix the conditions of a co-operation project is spent talking business and 75 per cent finding the most sophisticated solution in order to avoid double and triple taxation of money transfers and profits.[24] The formulation of a common EC company law, towards which the first steps have been taken with the introduction of the European Economic Interest Group (EEIG) in 1989, is an indispensable prerequisite for multinational consortia that want to compete for arms contracts on a West European level. There are clear indications that the Commission will persist in its efforts to get control of the defence market. Currently two studies are being conducted by the Directorate General for Industrial Affairs. One focuses on the procurement of defence equipment in the Community, the other on the situation and prospects for the restructuring of European defence-related industries. In addition, the need for co-ordinated regulations for arms exports of EC members cannot be denied. It is not yet clear how this need will be met.

In the 1990s both frameworks for the West European arms market will probably coexist. In the longer run, however, the institutional weakness of the IEPG, in particular its dependence on EC regulations, could tip the balance in favour of the latter. It was reported that before the Copenhagen meeting of the IEPG in November 1990 the Dutch Defence Minister sent a letter to his colleagues pleading that the tasks of the IEPG be taken over by the EC. No decision was taken, pending the outcome of the Intergovernmental Conference on Political Union.[25]

[24] 'Défense Européenne—l'horizon 1992: actes du colloque', *Revue Passages* (Paris), special edition, 1989, p. 34.

[25] 'Intensiever contact IEPG en EG', *Europa van Morgen*, 28 Nov. 1990, p. 698.

4. Political change and world arms export markets: impact on the structure of West European arms industries

Frederic S. Pearson

I. Introduction

The various political and economic changes in Eastern and Central Europe portend much re-evaluation and restructuring of both strategic plans and military preparations in Western Europe. While only a portion of defence policy depends on security threat analysis (with the other segment accounted for by economic welfare concerns, entailing jobs, technological advancement, corporate profits and arms design capability),[1] the reality of Soviet military withdrawal, governmental reform and the dismantling of the Warsaw Treaty Organization (WTO) necessarily diminishes the impetus for NATO armament.

Yet in any of these scenarios for a new European security system, whether based on NATO or the Conference on Security and Co-operation in Europe (CSCE), defence forces would continue to have a role, and armaments would be required. Indeed it has been argued that while NATO arms standardization and procurement diminish, certain forms of European standardization and armament must increase, as in the development of command, control, communications and intelligence (C^3I) systems both for general defence and for necessary 'surveillance, inspection, communications and many other forms of supervision and verification to encourage confidence-building between the two power blocs'.[2] Thus, the extent of European armament might be far lower than that projected in the recent past, and force configurations might be designed more for passive than for active defence, or at least for reassurance as much as deterrence,[3] but the temptation to develop high-technology equipment for these needs would persist.

[1] See Kolodziej, E. A., *Making and Marketing Arms: The French Experience and its Implications for the International System* (Princeton University Press: Princeton, 1987); Kolodziej, E. A. and Pearson, F. S., 'The political economy of making and marketing arms: a test for the systemic imperatives of order and welfare', *Occasional Papers*, no. 8904 (Center for International Studies, University of Missouri–St Louis: St Louis, Mo., 1989).

[2] 'The European market for military C^3I', *FS Reports*, no. E1292/M (Frost and Sullivan: New York, winter 1990).

[3] See Mandelbaum, M., 'Reconstructing the European security order', *Critical Issues*, no. 1 (1990) (Council on Foreign Relations: New York, 1990). Mandelbaum uses a distinction developed by Michael Howard.

Increasingly, the realities of limited national market size and escalating cost and investment requirements, not to mention the now diminishing NATO arms requirements, have threatened the myth of national arms autonomy. In the Federal Republic of Germany, France and the United Kingdom, arms companies have already largely been consolidated in each area of production (land, sea, air, electronics, engines, etc.), and resulting 'national champions' have been subsidized and nurtured.[4] The latest developments in the emerging European economic and currency market, in European political reform and in Third World, North American, and Asian markets, now appear to be pushing the existing arms production patterns into a more complete restructuring, which is likely even further to eclipse European national boundaries as mergers, buy-outs and multinational consortia develop.

The role of smaller EC members is also affected by these likely changes, as states make the choice of whether to join or refrain from further multinational arms projects, projects usually dominated by manufacturers from the larger advanced military powers. Neutral countries such as Sweden must contend with a tradition of defensive self-sufficiency in an age of rapidly increasing defence costs and receding threats.[5] East European states will enter the new strategic and economic dialogue within a 'common European home': arms production, although scheduled for cutback by 20 per cent in the USSR and on the decline in recent years in the rest of Central Europe,[6] remains traditionally among the most efficient and internationally competitive sectors of their economies.

The purpose of this analysis, then, is to predict how the overall evolution of the international arms market is likely to affect the European arms industry and its international ties. The first task is to determine which international political and economic changes are likely to affect arms production and procurement in general, and to gauge the effects of these in Europe.

II. The evolving market

Europe is not the only place in the world undergoing major recent transformations and political changes. Indeed, European developments rebound in various regions, particularly with Soviet and concomitant US withdrawal from Third World competition. Effects of greater superpower co-operation, or muted competition, have been seen from the Persian Gulf to Southern Africa and South-East Asia. Important superpower clients can no longer necessarily count on assured access to arms on concessionary terms through military assistance

[4] Moravcsik, A., 'The European armaments industry at the crossroads', *Survival*, vol. 32, no. 1 (Jan./Feb. 1990), pp. 65–85. See also Wulf, H., 'Arms production', SIPRI, *SIPRI Yearbook 1991: Armaments and Disarmament* (Oxford University Press: Oxford, 1991), pp. 281–316.

[5] See chapter 15 in this volume. See also Rapp, J., 'Sweden moves to aid troubled arms industry', *Jane's Defence Weekly*, 24 June 1989, p. 1338; Ljung, B., 'Under the gun: a survey of the defense industry', *Sweden Now*, no. 2 (1989), pp. 19–21.

[6] See Curtis, I., 'Europe's defense industry faces turbulent times' *Defense and Foreign Affairs*, vol. 18, no. 1 (Jan./Feb. 1990), p. 14; Matejka, Z., 'Building an all-European security system: a Czechoslovak view', *Disarmament*, vol. 14, no. 1 (1991), pp. 32–34.

programmes. As the Afghanistan conflict winds down, US decision-makers have been under greater pressure to justify high levels of arms transfers to Pakistan, for example. The USSR has enunciated a doctrine of 'reasonable defensive sufficiency' towards Syria, under which limits on sophisticated weapon transfer are outlined.[7] This reduces the political bargaining power of potential arms customers. On the other hand, their commercial bargaining power, albeit with persistent problems of credit, may be enhanced. Potential surpluses of military stocks could lead to increased superpower reliance on military as opposed to economic assistance programmes, as in US efforts to supply Peru's government for counter-insurgency and counter-drug warfare.[8]

Partly because of superpower co-operation, a number of lingering and destructive Third World military disputes and wars also are in a period of at least suspended hostilities. Again, South and South-East Asia, the Persian Gulf, Southern Africa, along with Central America, all have seen peace negotiations, if not outright settlement of persistent conflicts. The demand for weapons to continue these disputes has not disappeared, as demonstrated in ample supplies to the factions in Afghanistan, but the pace of arms demand, and with the Persian Gulf War even the impetus to supply, has slackened.[9] States located near chronic trouble-spots, such as the Persian Gulf, continue to arm, and wars flare periodically (as in Kuwait). In addition, there are plenty of breeding grounds for new wars given political disputes in the Middle East, South and South-East Asia, East Central Europe, the Pacific and Africa.

It is well-known that the last few years of the 1980s brought an overall decline to Third World arms sales, although regions such as the Middle East continue to import increasing quantities and values of armaments. With the suspension of the Iraq–Iran War, even Middle Eastern imports of major weapons were reduced by about 75 per cent. By 1989 North African imports had rebounded, but in South America and sub-Saharan Africa imports were down considerably as compared to 1984. The declining oil market diminished available purchasing power, foreign debts accumulated and weapon inventories were filled.[10] However, while strategic interest in supplying clients also may be diminishing with the passing cold war, new localized strategic concerns are being cited, at least in Washington, to justify both the continued production of

[7] These developments are outlined further in Pearson, F. S., 'Prospects of Third World arms transfers: patterns of demand and conflict', Proceedings of the Conference on Conventional Forces and Technology, organized by the Southern Methodist University and the University of Texas–Austin, Dallas, Dec. 1989 (forthcoming).

[8] Day, M. R., 'Peru balks at US military-aid offer', *Christian Science Monitor*, 3 May 1990, p. 3.

[9] See Grimmett, R. F., *Conventional Arms Transfers to the Third World, 1983–90*, Congressional Research Service Report no. 91-578 F (US Library of Congress: Washington, DC, 2 Aug. 1991), p. 6. On renewed moves to end conventional arms transfer controls, see Wit, J. S. and Spector, L. S., 'A realistic path to Mideast arms control', *Christian Science Monitor*, 13 June 1991, p. 19.

[10] Anthony, I., Courades Allebeck, A., Hagmeyer-Gaverus, G., Miggiano, P. and Wulf, H., 'The trade in major conventional weapons', SIPRI, *SIPRI Yearbook 1991: World Armaments and Disarmament* (Oxford University Press: Oxford, 1991), appendix 7A, pp. 228–31; Turner, J., SIPRI, *Arms in* the '80s*: New Developments in the Global Arms Race* (Taylor & Francis: London, 1985), pp. 13–15; Catrina, C., United Nations Institute for Disarmament Research (UNIDIR), *Arms Transfers and Dependence* (Taylor & Francis: New York, 1988); Grimmett (note 9), p. 46.

controversial weapons and the sales of previously restricted systems (such as US promotions of the M-1 Abrams tank to Saudi Arabia).

Although arms transfers to the Third World generally peaked between 1983 and 1984, continued fighting in the Middle East spurred a few more years of major exports in the latter half of the 1980s. The arms export record of the major world suppliers was very spotty during this period with a year of significant sales increases (frequently 1986-87) followed by three consecutive years of significant export decreases. In terms of major weapons, only Brazil, Israel, the UK and the USSR averaged an overall annual increase during the entire period 1984–89, and only for the first two was the average increase more than 5 per cent. However, Soviet exports dropped considerably in 1990, and Israel and Brazil experienced several arms production and sales difficulties by 1989, owing to capital shortages, debt and technological factors. Among the states with the greatest annual decline of sales during the period were France, the FRG, Italy and Spain. The USA experienced modest declines in sales of major weapons as well. Although the years 1985–87 saw sales increasing in the neighbourhood of 25 per cent annually, and new US sales agreements increasing markedly in 1986–88, a reduction can be seen in 1990.[11]

Rather than abandoning ongoing weapon system production in response to declining or intermittent demand, major suppliers have first explored increased export potentials, even including government subsidy of foreign purchases such as renewed US emphasis on Foreign Military Sales (FMS) agreements following a decade of stress on direct commercial sales. Similar British and French credit guarantees for Iraq proved quite embarrassing in 1990.[12] Despite lingering military concerns about weapons winding up in the 'wrong hands', there is greater major power willingness to sell excess and previously restricted weapons on the open market, especially to states able to pay cash for the products. This is due to concern about diminished NATO and WTO procurement needs and surplus production capacity. Yet the 'peace dividend' pressures experienced by a number of major powers as the cold war ended in 1989 are ultimately likely to cause extensive production and procurement cutbacks, which would reduce the arms volume available for export. At the same time, however, firms could be freer to produce more export-oriented designs.

The demand for technology transfers and technological co-operation is also a factor shaping the international arms market, and one which continues to gain importance. The production and acquisition of arms through imports and licensed production, as well as agreements on 'offset' economic inducements to purchase arms, are all seen as a form of technological 'quick fix' for governments in less developed countries (LDCs) hard pressed to provide jobs for engineers and trained personnel, albeit at the cost of displaced development

[11] Anthony, Courades Allebeck, Hagmeyer-Gaverus, Miggiano. and Wulf, (note 10), Table 7A.2, pp. 230–31.

[12] Isnard, J., 'Debt rescheduling frees French to sell to Iraq', *Jane's Defence Weekly*, 20 Sep. 1989, p. 674; Wilson, J. R., 'U.S. industry must look overseas', *Jane's Defence Weekly*, 16 Sep. 1989, p. 531; Hunter, S. T., 'Time to stand up to Iraq', *Christian Science Monitor*, 9 Apr. 1990.

capital.[13] Whether the development of an arms industry is the most efficient way to acquire technological sophistication is open to doubt, but it may be one of the few obvious options readily available to LDC leaders. It also fits their interest in security and trade. Even if the country in question does not have an immediate foe on its borders, independent arms production capacity provides a form of status that is expected to enhance international influence and thus security.[14]

Partly for these reasons, and to offset battlefield deficiencies and foreign dependencies, lethal weapon technologies have spread alarmingly in recent years. A number of LDCs are now able to produce, modify or gain access to relatively sophisticated missile and warhead delivery systems, and have received help from outside powers in developing these capabilities. Argentina and Brazil, for example, hard pressed by international debts, reportedly have been in a position to make their considerable 'dual-use' nuclear as well as missile technology available more widely to cash customers such as Algeria, Iraq and Libya. European technology is implicated in many of these deals as well.[15] States unable or unwilling to produce nuclear weapons are nevertheless tempted to obtain other weapons of mass destruction, such as chemical weapons. In fact, one of the strongest correlates of arms importation in the Third World is an evolving national capability to produce such lethal devices, indicating a general militarization of political goals.[16]

Despite such attempts to escape from dependency on foreign arms and technology, much of the Third World remains highly dependent on both. The USA and the USSR still account for up to 90 per cent of global military R&D. For certain advanced systems and components, such as engines (or clandestinely obtained sophisticated nuclear triggering devices), as well as for training and logistical support, demand for major power equipment and input is likely to remain very high, especially since major powers also maintain influence on other arms supplying states, such as Brazil and Israel, through licence arrangements, and since modern battlefield operations often require satellite intelligence. Growing debt burdens and economic downturns, heightened restrictions applied by major powers as well as differing motivations for producing arms (e.g., Israel's security–autonomy–trade interests compared to South Korea's mainly commercial production under US licence) still make it extremely difficult for Third World arms industries to advance

[13] On these effects and the debates surrounding them, see Mehta, J. S. (ed.), *Third World Militarization: A Challenge to Third World Diplomacy* (Lyndon B. Johnson School of Public Affairs, University of Texas: Austin, 1985); Kupchan, C. A., 'Defence spending and economic performance', *Survival*, vol. 31, no. 5 (Sep./Oct. 1989), pp. 447–61.

[14] Kolodziej and Pearson (note 1), with specific reference to Brazil.

[15] Ehrmann, E., 'Libya shops for nukes', *Christian Science Monitor*, 3 July 1990, p. 19.

[16] Pearson, F. S., 'The correlates of arms importation', *Journal of Peace Research*, vol. 26, no. 2 (May 1989), pp. 153–63.

much beyond the rudimentary production capabilities of relatively simple systems.[17]

III. Implications for the European arms industry

Even before the latest political and market changes emerged, the European arms industry was under some pressure to restructure and reorganize. The rising cost of weapon R&D meant that fundamental questions of market size and return on investment were confronted throughout the 1980s. With massive cost and price increases, only declining quantities of high-technology weapons could be produced with essentially fixed European defence and military R&D budgets. While the latter may now also diminish somewhat, the USA's 3 : 1 R&D advantage over Europe probably will not change much in the foreseeable future.

Military procurement priorities, as well as the high-technology race, brought the major manufacturers inexorably into the production of sophisticated weapons mainly for the NATO market. Along with improving 'two-way street' sales to the USA (closing the gap to merely a 3 : 1 advantage for the USA there as well), they were also able to market certain of these designs (mainly aircraft, tanks and missiles) quite successfully in the Third World, along with lighter speciality designs such as naval patrol craft (especially from France and Italy, with the FRG contributing in submarine design). International production consortia were established, often, however, without continuity from one project to another and usually excluding a key weapon producer, such as Dassault of France, whose products then competed with the joint project. Wealthy Third World customers, interested in the prestige and combat value of advanced weapons, were sought to fill in the orders for longer production runs. A close Euro-Arab/Middle Eastern connection was fostered, particularly to displace US arms when congressional limitations were imposed. Technical, bureaucratic and legal complexities all increased as such weapon systems and overseas sales were engineered.[18]

While these 'successes' were being registered, certain weapon systems were already being reconsidered or abandoned before the Berlin wall was breached. In some areas of technology, such as airborne radar, Europe is hard pressed to compete internationally for lack of a single viable manufacturer. The NATO Modular Stand-Off Weapon (MSOW) was abandoned due to high costs and failure to make a coherent weapon out of modules. On the other hand, the much heralded NFR-90 NATO frigate project evidently lost its major participants because of over-standardization and insufficient use of modular concepts. Thus,

[17] Neuman, S., 'Dependence, power and influence: the role of military assistance', ed. C. M. Brandt, *Military Assistance and Foreign Policy* (Air Force Institute of Technology, Wright-Patterson Air Base: Ohio, 1989), pp. 7–32.

[18] These included the need to involve numerous subcontractors and co-developers, and to obtain investment capital, payment (sometimes in terms of bartered goods such as oil) and market guarantees. See Curtis (note 6).

the competitive sufficiency of certain European weaponry and joint weapon projects remained questionable.

Now that the impetus for joint NATO procurement itself is slowing, and with the 1992 unification envisaged in the EC Single European Act, these trends and questions are magnified. Justification for cost-reducing longer production runs and expanded military R&D will become increasingly difficult. Both the search for non-NATO markets and the need for flexibly designed, light, multi-role weapons to supply them will grow. Indeed, the production of primarily export-oriented designs would involve competition with smaller non-European arms exporters as well as with the USA and the USSR, and would be another reversion to the patterns of the 1930s, when manufacturers such as Vickers engineered whole product lines which were never procured by British forces but were sold instead piecemeal abroad to countries such as Chile, Finland and Thailand.[19]

None the less, the West European tradition of advanced major weapon systems, heavily oriented to major power warfare, will die hard, both for reasons of governmental, military and corporate inertia, and of ongoing ties to demanding customers such as those in the Persian Gulf.[20] Just as it was hoped that the expanded Euro-market of 1992 would spur economies of scale sufficiently to generate high-technology consumer products to compete with Japan and the USA, a significant body of European opinion harboured the same hope for military technology, even in the midst of declining real defence budgets and despite provisions by which the Treaty of Rome exempts the defence sector. Many of these hopes have been pinned, at least in the medium term, on heavy basic research in multi-use technologies under the EC Framework Programme and its R&D and technical co-operation programme, Eureka.[21]

However, if changing economic and security incentives lead to a merged West European arms market, massive over-capacity and duplication of production in many systems will be exposed, with resulting intense interfirm competition and 'merger mania' (conceivably involving government subsidies) and/or production constraints.[22]

European arms production has hitherto been stratified between the three major arms producing states (France, the FRG and the UK) on the one hand and on the other the smaller states and manufacturers who respond partly as

[19] Curtis (note 6), p. 14. See also Kasslow, A., 'Defense industries seek ways to adjust', *Christian Science Monitor*, 30 Apr. 1990. p. 3.

[20] While warfare in much of the Third World probably will remain relatively localized and limited, the 'electronic battlefield' will increasingly be a global phenomenon. Even 'light' weapons will increasingly involve sophistication originally designed for the Central European front, and will find ready markets among wealthier non-European customers. On the continuing appeal of 'big ticket' products in the European defence industry, see Curtis (note 6), p. 14.

[21] Pearson, F. S., 'European security policy and the Single European Act', Paper presented at the Conference on the 1992 Project and the European Community, Florida State University, Tallahassee, Apr. 1990; Hitchens, T., 'Europeans seek to strengthen high technology role', *Defense News*, 16 July 1990, p. 8.

[22] Barrie. D., 'Euro-industry faces brutal restructuring', *Jane's Defence Weekly*, 7 Oct. 1989, p. 741; Moravcsik (note 4), pp. 67–68.

subcontractors to the main producers and partly with specialized designs. The bulk of the business has entailed the sometimes redundant and sometimes co-ordinated production of the big three, occasionally augmented by Italy. Sweden has been a major producer somewhat outside the West European network.

The British and French have been roughly equivalent in arms output, both in terms of numbers of firms among the (Organization for Economic Co-operation and Development (OECD) top 100 companies, and in terms of these firms' turnover. Moreover, the British and French firms, along with those of Italy and the Netherlands, are among the largest in average sales per firm, thus roughly equal in average firm size. Swedish firms are also among the world's largest, but on average produce only half to one-third the sales volume of the 'big four' (see appendix D). However, another study reports that strictly among the top 100 European arms firms, Britain has a 2 : 1 advantage (30 : 15) in the number of arms firms over both France and the FRG, with Italy a close fourth.[23] Total sales of the 41 major European arms firms are roughly $52 billion, which is the lion's share of the estimated $70 billion in sales for the 100 largest Euro-arms companies.[24]

Thus it can be argued that both the UK and France remain highly dependent upon the arms trade, but in different ways. Both are subject to considerable economic dislocation with a constricting NATO market and uncertain Third World sales prospects. Britain has more firms at risk, a slightly larger overall arms sales volume, and a greater traditional stake in NATO procurements. However, France is more vulnerable in terms of dependency on foreign arms exports as a percentage of production, with markets concentrated on fewer products from fewer firms and exported to fewer countries. British arms contractors also generally appear to have less of their overall business engaged in arms production (ranging from 10 to 45 per cent) than French contractors (an average of 59 per cent and median of 63 per cent among the 24 major arms contractors in 1987).[25] However, these comparisons of stakes in arms exports depend in part on the weapon in question. It has been argued, for example, that French tank manufacturers are less export dependent than their British or German counterparts, partly because the domestic French tank market, committed to the purchase of the Leclerc (from 1992), is more secure.[26]

Weapon manufacturing firms in much of the industrialized world have three basic options to confront current changing economic and political conditions. They can abandon the arms business largely or entirely rather than face uncertainties over government subsidy and procurement and possible contract cancellation. Concomitantly, they can contract arms production to a small and focused segment of their business and hope for a niche in the market. Or they

[23] Curtis (note 6), pp. 17–20.

[24] Curtis (note 6), pp. 17–20. Curtis also estimates that the top 100 European countries employ roughly 516 000 people in defence-related work, a low figure by other counts.

[25] Branger, J.-G., *Avis présenté au nom de la commission de la défense nationale et des forces armées sur les projet de loi finances pour 1989, Tome V, Défense, recherche et industrie d'armement* (Assemblé Nationale: Paris, 13 Oct. 1988).

[26] Baker, C., 'UK buy key to survival for tank makers', *Defense News*, 16 July 1990, p. 4.

can expand, absorb rival companies and seek to compete on a grand scale against other large conglomerates. In a sense, the last option has been exercised for many years within European states by both governments and companies promoting consolidations and mergers to remain competitive. This has resulted in 'national champions' such as British Aerospace, and more recently Daimler-Benz-MBB (see chapters 7, 10 and 11).

With the Single European Act, the way is eased to extend such mergers across national boundaries, in effect to create 'Euro-champions', although evidently few arms firms have taken advantage of the changing rules as yet. Aside from cross-national mergers *per se*, production consortia, such as that involved in the Tornado fighter programme and that envisioned by some for the next generation battle tank,[27] also fit the priorities set by the Inter-European Programme Group (IEPG), enunciated in 1988 (see chapter 3 and appendix F).

Despite progress on these fronts, intra-European arms trade has yet to expand greatly. In the search for cost efficiencies, the leading European countries have been more receptive to foreign purchases and partnerships than have countries with smaller and less efficient industries.[28] However, even these hesitant moves towards European-wide weapon procurement are coming at the cost of high offset payments (up to 100 per cent of the price), such as those offered by French and German as well as US tank manufacturers seeking to compete with Vickers to replace the UK's old Chieftain tanks.[29] With significant sales competition and offset requirements even inside NATO (the political future of which is itself uncertain), and with declining needs and government funds for many new weapons, prospects for greatly expanded, harmonious joint weapon production or procurement remain problematic.

Fears exist as well that the impetus towards a merged arms market might not necessarily result in equipment that is easily marketable outside Europe, and might swamp the arms production and options of smaller EC members and neutrals. Until now consortia have been developed mainly to compete for NATO sales and to expand joint procurement enough to make the products cost-effective. Some of these products (e.g., the Tornado aircraft) have been marketed to richer Third World customers, sometimes in lieu of US products, but other products (e.g., helicopters) have had less global market success. Even successes such as the Tornado later ran into cost problems in the complex barter arrangements with customers, as in the case of Saudi Arabia. Often it has been simpler to market designs (such as for naval patrol vessels) devoted mainly to export and less in conjunction with NATO priorities. Therefore, the question of what will be the ultimate main market for and characteristics of European weapons in the coming decades remains open.

[27] Baker (note 26).

[28] Curtis (note 6), p. 12.

[29] Miller, C., 'Krauss Maffei offer 100 percent offset deal in UK tank purchase', *Defense News*, 9 Apr. 1990, p. 3. The incentive here is not only the British market, but further expected export openings to clients such as Canada, Kuwait, Malaysia, Morocco, Pakistan, Saudi Arabia, Sweden and the United Arab Emirates. See Baker (note 26).

Table 4.1. Likely stratification of the European arms economy in the 1990s

Stratification	Orientation
1. Type of product	*Market*
Simplified/specialized	Mainly extra-European/Third World
High-technology	Europe, the Americas, advanced Asia and the Middle East
2. Orientation of international partnerships	*Activity*
European mergers and acquisitions	Co-development
European consortia of producers	Co-development
Euro-American–Asian consortia	Co-production/co-development
Europe–LDC[a] licence agreements	Co-production (gradual co-development)
3. Type of co-operative/competitive structure	*Type of weapon*
Multinational co-development cartel(s)	Large systems ($2–5 billion)
Competing consortia	Medium-sized systems ($0.5–2 billion)
Managed inter-firm trade	Smaller components and low-technology systems (less than $0.5 billion)

[a] Less-developed countries.

Source: Adapted from Moravcsik, A.. 'The European armaments industry at the crossroads', *Survival*, vol. 32, no. 1 (Jan./Feb. 1990), p. 77.

In addition to the existing hierarchy of weapon manufacturing states, therefore, three basic forms of stratification appear likely to characterize future European arms production (see table 4.1):

1. Stratification by type of product lines;
2. Stratification by orientation of international partnerships;
3. Stratification by type of international co-operation and 'open market' incentives.

The first form means that some states and consortia will aim primarily at non-European markets with simplified and specialized export-oriented product designs. Judging by past records, this would be likely to include groups of smaller firms able to shift production priorities rather flexibly and to establish ties to LDC manufacturers, or those with relatively little of their business volume in arms production. Despite lingering political and structural barriers, Central and East European states could also join such projects, given their traditional access to certain key Third World markets and their traditionally well-honed product lines, which have produced various kinds of military equipment with reasonable efficiency in large quantities over many years (e.g., the AK-47 rifle, tanks and infantry equipment).

On the other hand, other firms and consortia can be expected to aim primarily at the more 'high-tech'-oriented European, North American and advanced Third World arms markets. Suppliers in this category would include the traditionally dominant states and companies, as well as advanced states heretofore largely excluded from NATO procurement for lack of 'club membership', namely, Austria, Sweden and Switzerland. Finland and other smaller European states, such as Belgium, Greece, the Netherlands, Norway, Spain and Turkey, could dabble in such projects occasionally, normally playing subcontracting and subsidiary roles.[30]

In this connection, the French Foreign Ministry has issued a report advocating arms industry concentration on high-technology sectors with export potential. Accordingly, less competitive sectors with limited export potential, such as artillery and shell manufacturing, should be abandoned in favour of more efficient foreign suppliers. With such an approach, 'One or two industrial companies will stay in certain sectors, but others will either have to abandon entire branches, or have to give up prime contracting by finding co-operation abroad'.[31] Similar calls for massive restructuring have been heard across Europe. For countries such as France, this restrucuring would represent a revolution in security policy, by significantly increasing dependence on outside suppliers and eschewing across-the-board production.

When arms producers establish high-technology international arms consortia, they can go in a number of geographical directions, hence the second form of stratification. Even as intra-European consortia are developed, states heavily invested in the traditional North Atlantic arms trade, for example, Belgium, the Netherlands, Norway and the UK, might be likely to enter consortia or engineer mergers with North American producers (the latter presumably also opening up to partnerships with producers in Japan, South Korea and other Pacific countries). The FRG, Norway and the UK, among others, also could join with European Free Trade Association (EFTA) states, such as Austria and Sweden. Growing French–German strategic consultations, and the interest of both in establishing links to the East, point towards ties with states such as Czechoslovakia, Hungary, Poland and even the USSR. This would augment the French–German–Italian arms consortia that have already emerged in the helicopter field. Each of these groupings would offer new partners access to certain traditional Third World clients.

Despite the ultimate decline in arms production in the former German Democratic Republic, GDR resources offer the unified FRG a number of advantages in arms production and exports (as seen in the dispatch, reported in the German press, of weapons to the Middle East during the Persian Gulf War). German firms also enjoy newly emerging aerospace partnerships with Western and Asian firms. For example, rumoured Daimler–Mitsubishi joint aerospace projects would provide Japan with an alternative to reliance on somewhat reluctant US firms such as Boeing and General Dynamics for dual-use aerospace

[30] On Belgian priorities and dilemmas, see *Jane's Defence Weekly*, 3 June 1989, p. 1101; Lewis, J. A. C., 'Belgian PRB looks to European alliance', *Jane's Defence Weekly*, 1 July 1989, p. 1385.

[31] The rationale even for continuing the controversial French naval fighter, the Rafale, as well as most naval construction is questioned. See Davidson, I., 'France's defence "must go high tech"', *Financial Times*, 12 Mar. 1990.

technology and project co-development (in addition to offering Japan marketing access to Europe; Daimler would gain similar benefits in Asia). Political incentives emerge as Japan would be less susceptible to US pressure regarding its developing jet fighter and satellite industries. Yet prospects for the success of such endeavours remain uncertain given distances, diverse marketing traditions and the fact that Daimler is new to aerospace itself and Mitsubishi is not yet a major actor.

The third form of arms industry stratification in a sense subsumes the previous two, and involves the market structure thought to maximize arms production efficiency in projects of varying magnitude. Moravcsik proposes that co-development cartels, with major governmental subsidization and *juste retour* expectations, will be required to produce the most sophisticated technologies and the larger weapon systems (in the $2–5 billion development range—e.g., fighter aircraft and engines, large missiles, nuclear systems and helicopters). Here it is likely that major US firms will bid to preclude Soviet participation in Euro-partnerships. With vast US technological experience and advantages, as the Westlands Helicopter case showed, it is difficult to see how such bids could be resisted unless European states, including the UK, were determined jointly to hone their own technology.[32]

Moravcsik's second strata of arms sophistication, medium-level weapon systems (with costs ranging from $500 million to $2 billion, and including electronics, armour, smaller missiles, etc.), could be based on competing production consortia, with more open market competition and return on investment. It is here that certain European suppliers could have significant competitive advantages *vis-à-vis* Third World and North American suppliers (as in the case of German tanks), and also here greater competition among groups of firms would be possible. Soviet and East European participation could emerge, since Moscow already has considerable experience with competing design laboratories in the state-owned sector (MiG versus Tupolev versus Antonov versus Ilyushin versus Yakovlev, etc.). The USA could also fit in but would be less likely to dominate technologically at this level.

Finally, for smaller components and low-technology systems below the $500 million mark (such as transport planes and trainers, munitions and small arms, etc.), Moravcsik suggests that small firms and national firms could still compete in a form of 'managed free trade', familiar in today's European arms system. Indeed, when speaking of components it is well to remember that like the automobile trade, the global arms trade appears to be undergoing a structural revolution both in terms of modularization and multinational parts supply. It has been estimated that 30 per cent of the vaunted Euro-Tornado aircraft is composed of parts manufactured in the USA.[33] If present trends continue, and in the high-technology area it is difficult to see how they cannot, aircraft engines,

[32] Moravcsik (note 4), p. 77. If Europe were to try to erect such barriers, almost exclusive reliance would be placed on the UK, France and the FRG, which in addition to difficulties in competing with the USA would stir resentment in smaller EC states such as Denmark against 'military–industrial complexes'.

[33] Neuman, S., 'The arms market: who's on top?', *Orbis*, vol. 33, no. 4 (autumn 1989), p. 526.

radars, avionics and other units to build into or hang on to weapon platforms, will increasingly be internationally derived, and if necessary manufactured in relatively cheap LDC labour markets. This means that it will be difficult to categorize any weapon system as 'European' in origin, and that company management will think less of end-products and more of component supply opportunities. Current restrictions on end-use and customer destinations will also be progressively more difficult to enforce, unless an effective EC or United Nations reporting system is devised.

Firms are on notice that survival in the developing international arms economy will be increasingly difficult. Those that fail to diversify product lines and to anticipate and satisfy the demand for components or finished products in large numbers will have an especially difficult time. Those that overcommit themselves to arms production will have little or nothing to fall back on in the way of alternative civilian goods when markets vanish or diminish below the break-even point. Even among traditional prime contractors, those unable to bid flexibly for subcontracting roles are likely to starve for contracts under conditions of uncertain government subsidies. Those with better access to commercial credit and bond markets will be able to raise sufficient capital to compete at least at the third, lower-technology level of production.

IV. Conclusions

Changing international arms market conditions are making it more difficult for arms suppliers in all corners of the world. The USA and the USSR will be 'unloading' surplus arms supplies on Third World markets, narrowing the room for European suppliers.[34] Arms suppliers in newly industrializing countries, suffering from shortages of capital and technology, fail to move beyond relatively rudimentary systems. This lessens the competition for Europe, but at the same time, arms markets are constricting as Third World customers have an increasingly difficult time coming up with hard currencies or credits to purchase arms (even oil-rich states seek barter and credit arrangements, although this could again reverse with a bolstered oil market in the late 1990s). European arms manufacturers, perhaps more than any others in the world, are confronted with the difficulties of remaining viable producers of high-technology weapons. When predicting behaviour in the international arms transfer 'system', it is important to note historical trends. Arms marketing patterns are strongly correlated to power status. While also interested in commercial advantages, the USA and the USSR have for 40 years been the most strategically oriented of the major arms suppliers, shipping most of their weapons to favoured clients in strategic locations (Cuba, Iran, Israel, Pakistan, Saudi Arabia, Syria, Viet Nam, etc.). Indeed, since World War II, while keenly

[34] In the past Washington has, under certain circumstances, supported the defence exports of key NATO allies such as the UK and Italy, and may continue to allow US foreign assistance funds to be used for such purposes. See Neuman (note 33), pp. 525–26.

interested in commercial advantage, Washington has been more prone than other advanced economic powers to politicize all forms of trade and to glorify or exult in weapon technology *per se*. European arms exports have, since the 1960s, been driven by commercial interests, although in the case of France, the FRG and the UK, governments also strive to achieve an underlying interest in greater politico-military autonomy and occasionally, as with France in Africa, political influence. Europeans have been less prone to place embargoes on or restrict supplies to unfavoured states.[35]

However, as indicated by the EC trade sanctions against Iraq and Libya and by EC policy during the UK–Falklands/Malvinas War, power and political shifts may subtly reorient the arms transfer policies both of the superpowers and of the European states. Depending on the extent of nationalistic upheaval in central and southern Europe and in the USSR, the reduction in cold war tensions will *likely* mean fewer superpower controls on military transfers to favour clients (as in the Persian Gulf), alongside efforts to prevent the rise of new 'Saddam Husseins'. Selective, commercially driven restrictions on advanced technology, of the type established by the Co-ordinating Committee on Export Controls (COCOM), could increase, as demonstrated by the struggle of US aerospace firms to restrict Japanese access to US-derived technologies, while still seeking to enter co-development consortia with, and obtain technological hints from Japanese firms. Eroding US politico-economic and Soviet political hegemony also means that the superpowers will increasingly scramble for commercial (balance of trade) advantages through arms as well as other types of trade.

Concomitantly, if—and it is a big if, given the difficulties encountered thus far in trying to establish European-wide arms consortia and common European defence procurement and policy-planning—Europe develops a unified arms production and procurement 'community', along the lines envisioned by the IEPG, then one also might expect *increased* politicization of European arms sales. In other words, as political co-operation in the EC extends to the security and defence fields, the EC foreign policy machinery will begin acting more as a single force. Again, such extension of co-operation is by no means certain as East meets West (and neutrals) in Europe, but it would be to disregard history not to expect behavioural policy changes when power hierarchies shift.

This raises the possibility of establishing some sort of order out of the 'diffuse economic competition'[36] likely to characterize the coming stratified arms transfer system outlined above. Will suppliers and supplier consortia compete in a helter-skelter search for presumably shrinking global, if not regional markets, or will the call for renewed negotiations on 'rationalizing' the

[35] See Kolodziej (note 1); Pearson, F. S., 'Problems and prospects of arms transfer limitations among second-tier suppliers: the cases of France, the United Kingdom and the Federal Republic of Germany', ed. T. Ohlsson, SIPRI, *Arms Transfer Limitations and Third World Security* (Oxford University Press: Oxford, 1988), pp. 126–56.

[36] Louscher, D. J., 'International arms procurement in the 1990s', Paper presented at the Annual Meeting of the International Studies Association, Washington, DC, Apr. 1990.

arms market be heeded? The aim of such negotiations would be to establish: (*a*) legitimate and reliable security and technological transfers to customers in various regions; (*b*) stable expectations of market shares for arms suppliers' planning and investment purposes; and (*c*) dampened political tensions and arms races wherever possible? For the first time since 1971, there is a prospect for a renewed series of Conventional Arms Transfer Talks (CATT), this time broadened to include middle- and small-level suppliers and regionally prominent customers, particularly in the Middle East.

Prospects in this regard, as in expectations of a unified European arms market, are contradictory. The superpower competition and suspicion, if not the domestic political disputes which doomed the original CATT, have now largely disappeared.[37] Nevertheless, European states will have to struggle to find foreign arms markets, especially if the USA is not willing to increase its European arms imports. Arms manufacturers will be pressed to form workable consortia while retaining or developing important technological advantages. R&D investments will be very difficult, even for single favoured consortia, unless some ready markets are assured.

In noting the possibilities, a number of questions emerge. Would one set of manufacturers be content to allow others greater access to the more lucrative markets (West, South and South-East Asia)? Would states with traditional access to various buyers be willing to share such sales with other less advantaged arms producers? As the arms production system becomes increasingly internationalized and dominated by component manufacturers, such questions could become moot, yet France seems intent on retaining distinct national firms in at least the efficient higher-technology arms fields. By the same token, would arms purchasers be willing to restrict their market (and hence price) choices by aligning with certain suppliers?

Such questions could be approached by a system of subsidized pricing, and by credit and security guarantees from the major powers. Most arms buyers have relatively modest security needs and economic means, and could be satisfied by such a security regime. The major arms customers would be more likely to resist, as is the case of the nuclear non-proliferation regime. But even these states are experiencing the type of debt and purchasing-power problems that could make the difficult CATT regime bargaining at least feasible. Clearly, with the international arms supply system in flux and international tensions somewhat diminished, there has never been a better moment to approach the problem.

[37] On the failure of CATT see, Husbands, J. L. and Cahn, A. H., 'The Conventional Arms Transfers Talks: an experiment in mutual arms trade restraint', in Ohlsson (note 35), pp. 110–25.

5. Responses of the US arms industry

Judith Reppy

I. Introduction

The United States and Western Europe have long been engaged in a dual relationship at the systemic level.[1] They are military allies in NATO, with shared security concerns and co-operative arrangements for addressing those concerns, while in the economic sphere they are rivals, vying for competitive advantage and market shares. The tension between these two different relationships permeates the transactions surrounding the development and procurement of new weapons, where the logic of security co-operation directly confronts the economic self-interest of the various national arms industries. Debates over weapon standardization, the scope and membership of collaborative projects, the role of military technology in civil industry, rules for technology transfer, and the two-way street in arms trade illustrate the complexity of the relationship between the arms industries of NATO member countries.

For many years US–West European co-operation in arms procurement was characterized by a largely one-way flow of US technology and weapons to Europe. Even after the European members of NATO succeeded in rebuilding their arms industries, the balance in the arms trade continued to favour the USA, which enjoyed a cost advantage because of its large domestic market. Over time, West European governments have sought to redress this imbalance through collaborative projects and moves towards a unified West European arms market. The USA, not surprisingly, has seen these moves as protectionist, although the US arms market is itself protected by 'buy American' clauses and the tendency of military officials to prefer domestic companies over foreign bidders.

Profound changes are now reshaping the US and European arms market. The political upheavals in Eastern Europe have overturned the strategic assumptions of the last 40 years: military force structures and strategy must be rethought, with uncertain implications for arms procurement. It is already clear that budgets will be cut substantially. The prospect of reduced budgets comes on top of long-standing trends of rising costs for new weapons, shortened production runs and increasing competition in traditional export markets, trends that have already forced changes in the arms industry. In the USA there is additional

[1] The same could be said for US–Japanese relations. Although this chapter concentrates on the US–European relationship, much of US policy must be understood in the context of the perceived economic threat from Japan.

pressure on military spending from the need to reduce the federal deficit, a need that provides a powerful impetus for cutting defence spending independent of the recent improvements in the international scene.

The budget deficit is not the only economic problem facing the USA. US military strength is no longer matched by its economic performance. Productivity growth rates have lagged behind those of its principal competitors in Western Europe and Japan. The USA has lost its leadership to Japan in important technologies and has since 1982 had a negative trade balance in manufactured goods. The arms sector is not immune to these trends, and it has become dependent on imports of some Japanese components.[2] The loss of economic power has weakened US ability to influence world events, whether through aid to the reconstruction of East European economies or trade with Japan and Korea. However, there is no consensus on the best way to address these problems, but rather a deep division between those who believe that the government should intervene to bolster the US technology base, and those—well-represented in the Bush Administration—who adhere to a free market philosophy.[3]

What will be the response of the US Government and the arms industry to these changes in the strategic environment, both military and economic? The probable outcomes can be analysed in terms of changes in the policy networks that make up the military–industrial complex. The behaviour of private firms is likely to determine the new structures because of the entrenched opposition in the USA to government intervention in the market. New combinations of contractors will form as the size of the industry shrinks and international collaboration becomes more common. The increasing importance of foreign sources and multinational companies will limit the ability of the government to manage the defence industrial base. Government support of the technology base is under threat as defence spending decreases. However, unlike the European arms market, the size of the US arms market is likely to remain large enough to support an indigenous industry across all major programmes.

II. The US arms industry

The most significant aspect of the US arms industry compared to its European counterparts is its size. In the late 1980s, US total military spending was more than double that of all of NATO Europe; spending for equipment was also more than twice as great; and even arms exports, which are relatively less important to the US industry, were over 75 per cent larger in absolute terms. The US preponderance in spending over individual NATO countries is even greater. These figures go a long way towards explaining the differences between the US

[2] Defense Science Board, *Report of the Defense Science Board Task Force on Defense Semiconductor Dependency* (US Department of Defense: Washington, DC, Feb. 1987).

[3] See, for example, the divergent opinions expressed in Twentieth Century Fund, Task Force on the Future of American Trade Policy, *The Free Trade Debate: Reports of the Twentieth Century Fund Task Force on the Future of American Trade Policy* (Priority Press: New York, 1989).

and European arms industries and illuminating the forces propelling the European industries towards collaboration. The larger US market has supported more than one producer in most areas and provided a measure of competition without recourse to foreign suppliers.

The special features of the US arms industry are well-known.[4] Viewed analytically, the arms market is a market for specific assets; that is, the weapons purchased lack the anonymity of classical competitive goods.[5] Rather than being purchased off the shelf, their characteristics and prices are determined through individual, negotiated contracts between the military services and private contractors. Transactions are governed by regulations and informal 'rules of the game' that have developed over time, instead of by impersonal market forces. Examples of these rules are the practice of bidding low on contracts in the expectation of 'getting well later'; the practice of stretching out programmes rather than cancelling them when funding is tight; and the 'gold-plating' of basic requirements that occurs by introducing new and more sophisticated demands after a programme has been approved. Regulations that directly affect international collaboration are the rules governing technology transfer, offset arrangements for export sales and 'buy America' clauses. New, informal rules are evolving around international teaming arrangements as a guarantee of access to foreign markets.

The institutions involved in the US arms market are commonly described in terms of a military–industrial complex (MIC), comprised of the arms contractors, the project offices of the military services (which are responsible for the development and production of new weapons), the support staff in military headquarters and in the Office of the Secretary of Defense, and the relevant committees and subcommittees of Congress. It is useful to think of the MIC as a set of interlocking networks organized around specific weapon development projects and bound together in dependent relationships.[6] The contractors depend on the project offices for income and profits, while the military services are dependent on the contractors to develop and produce the equipment they need. A similar set of relationships binds subcontractors and contractors together. Congress is ultimately the source of the Department of Defense (DOD) budget, and individual members of Congress may be dependent on defence projects for employment in their districts.

Defence networks are also connected to other policy networks, above all to the networks centred on issues of industrial policy in general and technology policy in particular. The arms sector has played an important role in the forming of US technology policy. In part this is the result of the large resources

[4] Gansler, J., *Affording Defense* (MIT Press: Cambridge, Mass., 1989); McNaugher, T. L., *New Weapons, Old Politics* (Brookings Institution: Washington, DC, 1989).

[5] Yarbrough, B. V. and Yarbrough, R. M. 'International institutions and the new economics of organization', *International Organization*, vol. 11 (spring 1990), p. 246.

[6] Elzen, B., Enserink, B. and Smit, W. A., 'Military technology—unravelling the gordian knot', Paper presented at the Conference on Policies and Publics for Science and Technology, London, 7–11 Apr. 1990; Gummett, P. and Reppy, J., 'Military industrial networks and technical change in the new strategic environment', *Government and Opposition*, vol. 25, no. 3 (1990).

devoted to military research and development (R&D); during the Reagan Administration military spending for R&D rose to account for nearly 70 per cent of federal spending for R&D and around one-third of total R&D spending. Although the bulk of military R&D spending is devoted to the advanced and engineering development of specific weapon systems, which offer only limited scope for the spin-off of new technology to the civil sector, a small fraction (which is, however, significant in absolute terms) is allocated to more generic technologies. Furthermore, military demand has often been important in fostering the early development of advanced technology in the civil sector because the military have been willing to pay the high cost of first-generation devices. Examples range from the introduction of the manufacture of interchangeable parts in the first half of the 19th century to the development of semiconductors in the 1950s.[7]

Beyond its historical role as funder and customer of new technology, the DOD has taken on a more formal role in technology development in recent years, serving as a base for a series of government programmes to support dual-use technologies, such as the Very High Speed Integrated Circuits programme, the Sematech programme for semiconductor manufacturing technology and the high-definition television programme. These programmes, which have had the obvious but often undeclared purpose of bolstering the civil technology base and improving US industrial competitiveness, have been placed with the DOD because the Department has enjoyed ample funding and because the ideological bias against government intervention in the market place argued against placing them with a civil agency. The rhetoric of national security, which calls for a strong civil technology base as well as a defence industrial base, has been needed to justify government support for technology policy.

West European nations have not been indifferent to the potential advantage that US industry has enjoyed in civil markets as a result of the large government budget for military R&D and the allocation of part of it to the civil technology base. The debate over arms industrial policy in Europe and the USA owes much of its urgency from the perception that it is really civil competitiveness that is at stake. Although the extent of beneficial spin-off is a hotly contested policy issue,[8] West European nations have commonly credited US military spending with providing the USA with an effective technology policy, despite the absence of a formal framework for co-ordinating and implementing such a policy. For example, they felt sufficiently alarmed by the potential benefits to US industry of the Reagan Administration's Strategic Defense Initiative (SDI) to institute, under French leadership, their own co-operative technology programme, Eureka, to fund similar work for civil applications.

[7] Smith, M. R., 'Army ordnance and the "American system" of manufacturing, 1815–1861', and Misa, T. J., 'Military needs, commercial realities, and the development of the transistor, 1948–1958', ed. M. R. Smith, *Military Enterprise and Technological Change* (MIT Press: Cambridge, Mass., 1985), pp. 39–86 and 253–88, respectively.

[8] See the chapters by Albrecht, U., Reppy, J., Edgerton, D. and Kubbig, B., eds P. Gummett and J. Reppy, *The Relations between Defence and Civil Technologies* (Kluwer: Dordrecht, 1988).

III. The US response to the changing environment

Against this background, the central question of this chapter can now be restated: Given the scale and institutional structures of the US arms sector, what are the likely responses to spending cuts and the changes in Europe, political and economic? The advantage of using the language of networks to describe the military–industrial complex becomes apparent when the impact of these changes is considered. Because of the magnitude and apparent irreversibility of the changes, one would expect the US response to be expressed through substantial changes in the composition of the military–industrial networks and in behavioural norms (the 'rules of the game').

With shrinking budgets the membership of the MIC will shrink, and the relationship between the parts will shift. Any given project may involve foreign as well as domestic players, as the network changes to accommodate multinational teams of contractors, joint ventures or other collaborative arrangements. Mergers and acquisitions will eliminate some members of the network while giving other players greater power. Subcontractors and third-tier suppliers may be disadvantaged by foreign competition.[9] Contractors will no longer follow the standard 'rules of the game'; for example, they will not be willing to bid low on R&D contracts if more profitable production contracts are not likely to follow.[10] The relationship between the military and civil technology base will alter, as the military budget declines, and the Government's ability to influence the technology base may diminish with the larger role played by international transactions.

The proximate cause of these structural changes is the reduction in the defence budget that has already occurred and the expectation that the future will bring even larger cuts. In the near term, hardware programmes have been relatively protected compared to spending for manpower and readiness, and the standard practice of stretching out rather than cancelling programmes has been followed. The major aircraft review carried out in the spring of 1990, for example, resulted in decisions to reduce the number of aircraft procured in several programmes and to postpone production of the Advanced Technical Fighter (ATF), the new fighter plane currently under development; nothing was cancelled.[11] Maintaining investment in future weapons protects the claims of the armed services on missions associated with those weapons. Moreover, cuts in manpower and readiness take effect immediately, reducing budget outlays, whereas R&D and procurement accounts are spread out over a period of several years (thus the large backlog of orders from the high-spending 1980s).

[9] US Congress, Office of Technology Assessment, *Arming Our Allies: Cooperation and Competition in Defense Technology*, OTA-ISC-449 (US Government Printing Office: Washington, DC, 1990), p. 84.

[10] Bond, D. F., 'Defense, industry officials differ on steps needed to aid companies', *Aviation Week & Space Technology*, vol. 132 (28 May 1990), pp. 69–70, 72. See also McNaugher (note 4), pp. 155–57.

[11] The planned procurement of B-2 bombers was cut from 132 to 75, of A-12 Navy attack aircraft from 858 to 620, and of C-17 cargo aircraft from 210 to 120. See Morrocco, J. D.,'Cheney proposes stretchouts, cuts in B-2, C-17 programs', *Aviation Week & Space Technology*, vol. 132 (30 Apr. 1990), pp. 18–20. The A-12 was later cancelled as a result of cost overruns.

Will the Government choose to protect its funding of the defence technology base as well? It can be argued that investment in R&D is even more important in times of budget stringency, since it provides a hedge against future threats at a relatively low cost. Such arguments were made vigorously in the early 1970s as spending was cut following the Viet Nam War, and they are being heard again today. On the other hand, the technology base is sensitive to the composition of research funding, and funding for basic and applied research carries a low priority in military circles compared to programmes that address immediate military needs. Programme stretch-outs imply holding projects in advanced and engineering development, reducing the funds available within the R&D budget for the technology base.

There is a particular problem for technology base funding resulting from decisions taken during the Reagan Administration. Funds for SDI were placed in the technology base and contributed most of the growth in this category during the 1980s; they now constitute almost 45 per cent of the total.[12] Approximately one-fourth of the SDI funds are allocated to so-called 'critical technologies'. These are technologies that the DOD considers vital to future security needs, and they include several dual-use technologies, such as robotics, that are important to commercial competitiveness. There is a rhetorical commitment to protect these programmes, but also recognition that they are in a vulnerable position. Charles Herzfeld, former Defense Director for Research and Engineering, is thus quoted as saying that he would consider reprogramming funds to protect the critical technology programmes, but money for technology base programmes 'is the first thing to go when the SDI budget is cut'.[13] The technology base funding, including the critical technologies programmes, is extremely vulnerable to cuts in the SDI programme, cuts that are sure to come as Congress searches for ways to reduce total military spending.

Support for dual-use technology within the DOD budget suffers as well from widespread political opposition to industrial policy. The force of this objection became clear in April 1990 when Craig Fields was removed from the directorship of the Defense Advanced Research Projects Agency after approving funds for a joint venture with a manufacturer of gallium arsenide chips. In other actions, the Defense Manufacturing Board was terminated and DOD funds for the development of high-definition television were cut.[14] Despite the interest of some members of Congress, such as Senator Jeff Bingaman, Chairman of the Senate Armed Services Committee's Subcommittee on the Military Industry and Technology, in supporting the civil technology base, the Bush Administration remains opposed to anything that smacks of industrial policy.

[12] Gilmartin, P., 'Defense research budget to escape deep reductions', *Aviation Week & Space Technology*, vol. 132 (19 Mar. 1990), p. 59.

[13] Bond, D., 'Pentagon seeks to protect funding for critical technology development', *Aviation Week & Space Technology*, vol. 132 (16 Apr. 1990), pp. 63–64.

[14] Bond, D., 'Transfer of DARPA chief draws criticism', *Aviation Week & Space Technology*, vol. 132 (30 Apr. 1990), p. 26.

As the budget has tightened, the DOD has acted to improve conditions for its contractors by softening the regulatory regime. Rules that were introduced to increase competitive pressures and to improve standards of conduct in the aftermath of the 'waste, fraud and abuse' scandals of the early 1980s have quietly been dropped. A large number of individual procurement regulations has been eliminated in an attempt to reduce red tape, although much of the reduction can be attributed to the removal of duplicate regulations. The military services no longer require for all programmes that there are two companies capable of producing the same product ('second sourcing'); it is too expensive when the total number of systems to be procured is small. The 1980s shift to fixed-price contracts for R&D has been reversed, since contractors are no longer willing to accept the risks involved when there may be no production contracts to recoup lost profits.[15] Regulations governing technology exports to Eastern Europe, if not to the Soviet Union, have been liberalized, and it is likely that there will be changes in the US munitions list as well.[16]

US attitudes towards the European Community's market consolidation in 1992 are ambivalent. While greater European unity is welcomed, there is concern over the protectionist consequences of European industrial and agricultural policy.[17] The prospect of a common European market for arms procurement raises similar fears. A study by the Defense Policy Advisory Committee on Trade concluded that:

> The result could be a stronger, more competitive European defense industry similar in size and capability to the U.S. defense industry. A unified Europe may also become more protectionist. U.S. policymakers need to develop a strategy for preventing such protectionism, encouraging increased collaboration with European defense contractors and maintaining a competitive U.S. defense industry.[18]

As this quotation suggests, official policy remains firmly in favour of a liberal international order, indicating that the USA is not likely to adopt protectionist measures in the arms sector beyond those already in place.

Although individual firms see co-operative projects as a way to maintain access to the European market, from the Government's point of view the fading Soviet military threat has weakened the security rationale for sharing technology with allies. Thus, US Government policy towards collaboration with NATO countries may become less favourable.[19] Full-scale collaboration—that is, collaboration negotiated by governments and beginning with the R&D phase of a project—has proved difficult to achieve between the USA and Europe in any case, despite its usefulness as a way to promote standardization of military

[15] Smith, B. A., 'Contractors restructuring operations to reflect changing defense needs', *Aviation Week & Space Technology*, vol. 132 (28 May 1990), pp. 66, 68.

[16] Starr, B., 'DoD offers "core" control list', *Jane's Defence Weekly*, vol. 14 (14 July 1990), p. 40.

[17] Twentieth Century Fund (note 3), p. 11.

[18] US Defense Policy Advisory Committee for Trade Policy Matters, *A Report Outlining U.S. Government Policy Options Affecting Defense Trade and the U.S. Industrial Base* (DPACT: Washington, DC, Nov. 1988), p. 28.

[19] US Congress, OTA (note 9), p. 23.

equipment. In effect, the balance between the forces stimulating collaboration and the desire to protect national industries has been drawn differently by governments within Europe and across the Atlantic: a large internal market has meant that joint projects are not necessary for the USA, and European governments have chosen to use their resources to build European technology. The Nunn–Roth–Warner Amendment to the Defense Authorization Act of fiscal year 1986, which earmarked US funds for NATO co-operative R&D projects to be performed in the USA, has resulted in a number of joint programmes, but it is not clear how many will survive beyond the R&D phase, when special funds will no longer be available. Several co-operative programmes have already been cancelled as one participating government after another pulls out,[20] and the DOD target of allocating 25 per cent of its R&D budget to co-operative programmes by the year 2000 seems unlikely to be met.[21]

This list of policy implications of the changes in the arms market could be extended. It must be recognized, however, that there are serious limitations on the Government's ability to make policy in this area. It is not just the usual problem of reconciling conflicting policy goals—defence budget cuts may impinge on support for the technology base; competition policy may not be consistent with minimizing costs when programme purchases are small; and technology transfer to defence allies may undercut commercial competitiveness—although these remain important. Changes in the defence networks are making the application of US Government policy increasingly problematic. In particular, the increasing internationalization of the arms market has created US dependence on foreign suppliers in some areas, brought foreign ownership of some US defence assets and undermined the concept of a national defence industrial base. Moreover, individual firms within the industry are responding to changes in the arms market in ways that may not be consistent with the national interest, as defined by the Government. In particular, technology transferred through company-to-company deals may be used in the future against the USA, either literally, in weapons purchased by third countries, or figuratively, in competition for civil markets.[22]

The impact of the changes in the strategic environment falls unevenly across the arms industry. Not all contractors are affected equally by the budget cuts; each company will adapt a strategy based on its product mix and on its forecast of future military programmes in the area of its own technological strength. Thus, some companies are leaving the market entirely by selling off their arms subsidiaries—if they can (Chrysler has taken its arms subsidiary off the market because no suitable buyer came forth), while others are buying up competitors in order to solidify their market positions. US arms contractors diversified in the late 1980s, reducing their dependence on arms sales, but there is no evidence that they did so through conversion of defence facilities; instead

[20] US Congress, OTA (note 9), pp. 98–99.

[21] US Department of Defense, *Report of the Secretary of Defense Frank Carlucci to the Congress for FY 1990*, annual report (US Government Printing Office: Washington, DC, 1989), p. 68.

[22] US Congress, OTA (note 9), p. 24.

diversification is the result of cutting the military share of total production.[23] Past failures in attempts to compete in commercial markets seem to have convinced arms contractors that they could not succeed now. Major lay-offs are taking place throughout the aerospace industry, with more to come.[24]

The end result will be a smaller, more concentrated industry. Major arms contractors will be protected by their status as systems integrators and their monopoly on specialized, technological defence capabilities. From this position of relative strength the prime contractors have a great interest in building bridges into the European market, which offers an enlarged market in exchange for sharing technology and production and a higher rate of profit than they earn on DOD contracts. Subcontractors and third-tier producers, by contrast, are resisting the trend towards increasing internationalization of the arms market. Collaborative agreements open the US market to competition from European companies at the subcontract level, and offset arrangements guarantee production contracts to European producers. Thus, the very opportunities that have led European arms firms to seek alliances with US contractors are causing smaller US firms to call for protectionist measures, including bringing a court case against the DOD in order to try to block the waiver of 'buy American' clauses.[25]

These fears, however, must be kept in perspective. The US arms market is large by international standards and will remain so for the foreseeable future. The trend towards increasing international participation in the US market raises important policy questions with respect to the acceptable level of foreign dependency and the future of the technology base, but it is unlikely to erase the home market advantage of US producers when competing for US contracts.

IV. Conclusions

As defence budgets shrink everywhere and arms markets become increasingly international in character, the interesting question for US–European relations is whether protectionist or cost-sharing motives will dominate. The answer will have profound implications for the European arms industry in particular, but it will also affect the US industry. The argument in favour of international collaboration in order to share development costs and gain access to markets will continue, made stronger by falling budgets. In the USA the security rationale for sharing technology across the Atlantic has been weakened, but the interests of individual firms—US and European—may be best served by joining international teams. In this as in other policy areas US Government policy is subject to amendment by private action, all the more so because the whole area of industrial policy has been left to market forces.

[23] Chollick. E., 'FY '89 top 100 defense contractors: beyond the cold war', *CEP Research Report* (Council on Economic Priorities: New York, 1990); Smith (note 15).

[24] Wilson, J. R., 'How to beat the budget blues', *Jane's Defence Weekly*, vol. 13 (19 May 1990), pp. 980–81.

[25] US Congress, OTA (note 9), pp. 83–85.

US arms contractors have different strategies for coping with the shrinking arms market, but they all point towards increasing concentration in the industry and more international collaboration. The new policy networks that are the result of these changes amount to a new institutional context for arms procurement, one with altered behavioural patterns. There will be a difficult period of adjustment as government institutions not noted for their flexibility grapple with the implications of these changes for competition and investment in R&D and the defence industrial base.

The new situation may seem one of great difficulty and stress, but it should also be recognized as one of opportunity. After four decades of high military spending and international tension, the possibility exists for the evolution of a more modest military establishment in the USA on a scale appropriate to the new international realities. This is a challenge to be welcomed.

6. Science, technology and the peace dividend

*Philip Gummett and William Walker**

I. Introduction

There is a common presumption that the new political order in Europe and the concomitant easing of tensions between the United States and the Soviet Union will yield a peace dividend. That is to say, spending on military personnel and equipment is expected to fall, releasing resources for other uses. The defence sector has traditionally absorbed a disproportionate share of scientific and technological resources (people as well as money), at least in France, the United Kingdom, the USA and the USSR. Accordingly, questions arise about the consequences of the changed international security environment for national science and technology policies, and thereby for industrial competitiveness.

The impact of a peace dividend in Western Europe (and in the USA) relating to science and technology will depend upon at least the following factors: (*a*) future military requirements; (*b*) the characteristics of the technologies involved; (*c*) whether technological leadership lies with the defence or civil sector; (*d*) demand in the economy and the role of spin-off effects; and (*e*) policy choices made in industry and government.

This chapter reviews some of the more salient features of each of these factors. It will become clear that general statements about the effects of any peace dividend on science and technology should be avoided. A variety of influences is at work, giving rise to the prospect of different outcomes in different areas of science and technology and in different countries.

II. Future military requirements

To a considerable extent, future military requirements, and therefore the technological demands to which these will lead, will depend on the final outcome of the Conventional Armed Forces in Europe (CFE) Negotiation process, particularly of the phase that is expected to follow the first round of CFE.

More important than any formal arms control agreements, however, is the fact that the erstwhile political support for high levels of defence expenditure is waning in all countries, East and West. Moreover, the Western strategic and military arguments underpinning operational requirements, based on threat

* The authors acknowledge financial support from the Economic and Social Research Council/Science Policy Support Group programme of research into defence science and technology, and valuable comments from Judith Reppy and Steve Schofield.

assessments that have focused on scenarios envisioning an invasion from the East, have suffered a severe loss of coherence. That threat has now been greatly reduced, weakening the political consensus within NATO concerning flexible response. This has proven especially true in the case of short-range theatre nuclear forces, but can be expected to influence decisions on other weapon systems, as governments individually and collectively rethink their military roles and requirements.

In the UK, the Treasury is reported already to have used such arguments to block funding for work on an anti-tank weapon for the Harrier GR5 aircraft, and to have called for a review of the mix of land-based and air-delivered anti-armour weapons.[1] Another important test case has been the replacement for the British Army's main battle tank. The original expectation was that some 600 Chieftain tanks were to be replaced, but by mid-1990 it had become improbable that this assumption would hold. Intense campaigns for the order were launched by Vickers Defence Systems (UK), General Dynamics (USA), Krauss-Maffei (FRG) and GIAT (France), the last three in particular working with international groups of collaborators. In the event, and after a good performance by the Challenger 1 during the Persian Gulf War, an order for 130 Challenger 2 tanks was placed with Vickers in June 1991, together with a decision to upgrade 420 Challenger 1 tanks. Even more momentous may be future decisions about the European Fighter Aircraft (EFA) project. In the FRG, although a decision not to pull out of the development phase was made in June 1990, it looks increasingly unlikely that this policy will be pursued when a decision on procurement is due in 1993: the opposition Social Democratic Party is against continuation, and the Free Democratic Party, junior partner in the current coalition government, is also opposed.[2] Political uncertainty in the FRG may therefore drive Italy, Spain and the UK, if they wish to continue with the project, to reconsider its specification and scale, while if they look to exports for salvation they will encounter competition in the form of the already troubled French Rafale fighter and perhaps also the new US Advanced Tactical Fighter.

As well as arguments about cutting or altering existing projects, a debate may be expected about the equipment needs for the new roles that NATO collectively, and its members individually, may adopt. More specifically, this may involve increased capacity for out-of-area operations (and therefore naval power), for mobility within Europe, for better command, control and communications, for detection and verification equipment, and for upgrading the performance of existing equipment. As with previous arms control agreements, there will be pressure to acquire systems not limited by the CFE Treaty—most notably, in the case of CFE, missiles of all types. One may speculate that the USA and Europe will grow further apart, in terms of strategic interests and

[1] 'Treasury blocks MOD's anti-tank weapon update', *Financial Times*, 8 Nov. 1989; Alan Clark, Written answers to question, *Parliamentary Debates*, House of Commons Official Report, 6 July 1990 (Her Majesty's Stationery Office: London, 1990).

[2] Janssen Lok, J., 'Political test looms for EFA', *Jane's Defence Weekly*, 23 Dec. 1989, p. 1370; '"No" to immediate fighter pull-out', *Financial Times*, 29 June 1990.

associated equipment, with Europe being less concerned than the USA about the global reach of its military forces, although after the 1991 Persian Gulf War this is less certain. The growing interest in non-provocative defence will add a further dimension to this debate.

Greater emphasis will also be placed, especially by the smaller countries, on the upgrading of existing platforms, coupled with their being supplemented—and, in time, displaced—by lower-cost alternatives. There will thus be a shift to smaller platforms and to more advanced or 'smart' weaponry. Their attendant communications systems will exploit, where possible, the secular trend towards distributed information systems, thereby avoiding replacement of one form of technological centralization (the large weapon platform) by another (the central telecommunications switch). This trend would also help to fulfil the requirement for increased mobility in deploying forces to meet possible threats around the European periphery. An important, and insufficiently recognized, aspect of these developments will be an increased emphasis on electronic warfare (EW), counter-EW and 'soft kill' technologies.[3]

The implication for technology that may be drawn from these considerations is that demand for defence electronics and software skills will remain stronger than for mechanical engineering skills, except (as with helicopters) where these contribute to mobility.

III. Characteristics of the technologies involved

Casual reference to 'defence science and technology' suggests a false homogeneity of activity and a clear distinction from civil science and technology. The distinction is also implicit in the well-worn term 'spin-off', implying a sharp distinction between a civil and a defence sector, with the possibility of entities moving between the two, typically from the defence to the civil, but sometimes in the opposite direction.

There has long been a need for a better taxonomy of technologies.[4] When the variety of technologies is considered, it is clear that there is not so much a sharp distinction as a spectrum between civil and defence technologies. It is also clear that there is a hierarchy of technologies, beginning with materials and components at the bottom and running up to entire systems at the top. Even such a simple model as this indicates at least the need to move away from a simple dichotomy of civil and military technology and to talk instead in the plural,

[3] See the emphasis on these technologies, and also on the importance of missiles over platforms, in British Atlantic Committee, *Diminishing the Nuclear Threat: NATO's Defence and New Technology* (British Atlantic Committee: London, 1984).

[4] For one way of developing this idea, see Walker, W., Graham, M. and Harbor, B., 'From components to integrated systems: technological diversity and interactions between the military and civilian sectors', eds P. Gummett and J. Reppy, *The Relations Between Defence and Civil Technologies* (Kluwer: Dordrecht, 1988), pp. 17–37.

about civil and military technologies. Following Carter,[5] it may be noted that the strongest dual-use technology relationships are likely to be at the lower tiers of the defence suppliers, which raises a question about the capacity of prime contractors to convert systems skills to civil ends. Following Edmonds *et al.*,[6] it may also be noted that at least as much attention should be paid to possible changes in production technologies as to final defence products.

The implication is that the impact of the peace dividend upon science and technology will vary considerably from one area to another. Conversion may be expected to be easier the closer one gets to the generic technologies at the bottom of the hierarchy and (which is another way of saying the same thing) to the dual-use technologies. If development, rather than research, is cut (as may happen in countries which choose to maintain long-term defence technology bases at the price of shorter-term development programmes, which is what happened in the UK from 1945 to about 1949), then the more defence-specific activities will be cut, raising the problem of the redeployment of defence-specialist technical staff. If, on the other hand, research itself is cut, then the problems of redeployment should be easier.

IV. Does technological leadership lie with the defence or civil sector?

Historically, the 1990s will perhaps be seen as the decade in which the post-1945 presumption that defence technology demands higher performance standards than civil technology came to an end. As it turns out, this was an historically specific notion. Associated with it, at least in the USA, the UK and France, was an assumption that the generic underpinnings of defence technology should be developed through extensive dedicated defence research programmes.

During much of this century, industrial development was linked to the maturation of a range of technologies (specifically in the electrical, chemical and automobile industries) that had their origin in the last decades of the 19th century. Military conflicts were important in driving forward new technologies (above all in the electronics, nuclear and aerospace fields) where civil demands were still limited and the returns on research and development (R&D) investments very uncertain. Particularly with the advent of micro-electronics in the 1970s and 1980s, mass markets developed for high-technology products and changes in the generic technology base and in production processes began to occur mainly in response to civil demands. Moreover, the lead was taken by countries, notably Japan, that had little involvement in military technology. These developments lay behind a 1989 US Office of Technology Assessment

[5] Carter, A. B., 'Anatomy of the dual-use relationship', paper presented to the Workshop on Dual-Use Technologies: International Perspectives, Centre for Science and International Affairs, Harvard University, 21–22 Oct. 1988, p. 9.

[6] Edmonds, M., *et al.*, 'United Kingdom and United States dependence of foreign technology in defence research and development', *Science and Public Policy*, vol. 17, no. 3 (1990).

report,[7] as well as other US expressions of concern about US loss of leadership to Japan in a variety of military and civil technologies.[8]

The rapid rate of obsolescence in areas such as electronics also brought acute problems to the military sectors. The life-cycles of many weapon systems far exceeded those of new civil technologies (often decades against years), imbuing military systems with a technological 'drag' that made them appear increasingly backward. Moreover, the inefficiency of peacetime selection processes—when no large-scale wars are being fought there are no obvious criteria—meant that great difficulties were experienced in choosing among technological trajectories in a period of rapid technical change. As a result, many military technologies experienced cost-inflation linked to seemingly uncontrollable increases in complexity, while the unit costs of civil technologies tended to fall, often dramatically.

Two organizational consequences are worth noting. One is that military procurement agencies have tended to place greater responsibility on systems contractors for making technological choices, while also trying to expose them to greater financial risk in order to discourage unnecessary increases in technological complexity. The other is that the providers of systems and subsystems have become increasingly dependent on imports of components from firms whose primary markets are civilian, for instance in materials and semiconductors. Moreover, companies with strong civilian capabilities (such as in computing and software) have begun to threaten the traditional defence suppliers at several levels of the technological hierarchy. In the UK, for instance, a software company became an important prime contractor for naval communication systems.

As defence markets shrink, and civil markets continue to expand, these trends are likely to become more pronounced. In an important sense, the civil technological base is becoming more strategic than (while also less distinguishable from) the defence technology base, as the US scare about Japanese supremacy in semiconductors mentioned above illustrates. The implication over the longer term is that countries that are strong in civil technology, such as the FRG and Japan, could challenge the supremacy of the established powers in military technology if they so wished.

V. Demand in the economy and the role of spin-off effects

Spending on defence R&D has already begun to fall, at least in those countries (such as France, the UK and the USA) where it constitutes a large share of government or total national R&D spending. If the decline in spending continues and accelerates, and defence technologies become less distinct from civil ones, what are the likely consequences? The answer turns in part on

[7] Office of Technology Assessment, *Holding the Edge: Maintaining the Defense Technology Base*, report no. OTA-ISC-420 (US Government Printing Office: Washington, DC, 1989).

[8] See Riddell, P., 'US risks falling behind in new technologies', *Financial Times*, 23 May 1990.

beliefs about the significance of technological diffusion between the defence and civil sectors. The concepts of 'spin-off' and 'spin-in' are problematical, not least because they depend on an implied clear distinction between the military and civil sectors.[9]

Some commentators, citing examples in computer, nuclear energy, radar, jet engine and advanced materials (and also advanced project management skill) development, have claimed that the spin-off from defence-originated programmes has been economically beneficial.[10] They can be expected to argue that reduced defence R&D spending will mean less money for technological advance, and that therefore industrial competitiveness will be damaged unless governments fund equivalent 'quasi-military' projects, such as second-generation supersonic transports or transatmospheric vehicles.

Others argue that any such benefits can be achieved more cost-effectively. They note that countries with high levels of defence R&D spending (again, France, the UK and the USA) have enjoyed poorer overall economic performance than countries with low levels of spending in the past (such as Japan and the FRG), and argue that the human resources tied up in defence R&D and production depress the stock of highly skilled people available to the labour market.[11] Kaldor has suggested that the performance orientation of defence technology has led to 'baroque' embellishments of major defence technologies, largely regardless of cost.[12] This argument leads to the conclusion that the direction followed by defence technology has diverged from that of civil, thus making spin-off inherently unlikely.[13] Indeed, in the UK the Advisory Council on Science and Technology has estimated that less than 20 per cent of government-funded defence R&D is aimed at technologies that may have a potential for civil application.[14] However, this argument must be tempered by the conclusion of the House of Lords Select Committee on Science and Technology that the British defence R&D statistics are misleading, and that only about half of what is counted is truly R&D.[15]

[9] Albrecht, U., 'Spin-off: a fundamentalist approach', eds Gummett and Reppy (note 4), pp. 38–57.

[10] For examples of such arguments, see Vogel, D., 'Government industry relations in the United States: an overview', eds S. Wilks and M. Wright, *Comparative Government–Industry Relations* (Clarendon Press: Oxford, 1987), pp. 99–103.

[11] For a general review of the spin-off debate, see Weston, D. and Gummett, P., 'The economic impact of military R&D: hypotheses, evidence and verification', *Defense Analysis*, vol. 3, no.1 (1987), pp. 63–76; Reppy, J., 'Military R&D and the civilian economy', *Bulletin of the Atomic Scientists*, vol. 41, no. 9 (Oct. 1985), pp. 10–14; Kaldor, M., Sharp, M. and Walker, W., 'Industrial competitiveness and Britain's defence commitments', *Lloyd's Bank Review*, vol. 60, no. 162 (Oct. 1986), pp. 31–49.

[12] Kaldor, M., *The Baroque Arsenal* (Deutsch: London, 1982).

[13] Recently a number of analysts have sought to examine the historical record (showing, for example, the influence of the military on the development of work organization) or to review the relations between defence and civil technologies (including the conceptually difficult issue of dual-use technologies) from a variety of policy perspectives. See, for instance, Smith, M. R. (ed.), *Military Enterprise and Technological Change: Perspectives on the American Experience* (MIT Press: Cambridge, Mass., 1985); Mendelsohn, E., Smith, M. R. and Weingart, P. (eds), *Science, Technology and the Military: Sociology of the Sciences Yearbook* (Kluwer: Dordrecht, 1988); Gummett and Reppy (note 4).

[14] Advisory Council on Science and Technology, *Defence R&D: A National Resource* (Her Majesty's Stationery Office: London, 1989), paragraph 3.13.

[15] House of Lords Select Committee on Science and Technology, Session 1989–90, HL-44, *Research and Development Statistics* (Her Majesty's Stationery Office: London, 1990). See also Science Policy

It should also be noted that reductions in defence R&D spending will not necessarily lead to an increase in civil R&D. In most governments, there is no provision for automatic transfer of resources in this way. This is not how budgetary processes work. What will really determine whether reduced defence demand for skilled people will result in their transfer to civil R&D is the overall level of demand within an economy for scientific and technological skills. In a buoyant economy, investment capital is available for the development of good ideas, whatever their origin; and jobs at appropriate skill levels exist into which displaced defence scientists and technologists can move. In a stagnant or recessionary economy, neither of these conditions holds, and displaced people perforce move into work that does not draw directly upon their technical skills. The impact of a decline in volume of defence R&D can therefore be expected to vary between countries according to the level of general demand in their economies. The spin-off debate should be seen within the same context.

VI. Policy choices made in industry and government

In the end, the character of any peace dividend in science and technology will turn mainly on the choices made in government and industry in response to the new international security environment.

Since 1990 it is commonplace to note that the structure of the defence industries in Europe is changing.[16] The immediate trend is towards concentration and internationalization, either in the form of a variety of joint projects, strategic alliances and other kinds of 'teaming' or, in a new departure, through the formation of joint companies with parents in two countries, as in the cases of Aerospatiale and MBB in helicopters and Thomson-CSF and Ferranti International in sonars (see chapter 2 for details). This process can be expected to continue as competition for falling numbers of orders intensifies. There are also signs of transatlantic teaming (for example, the proposed link between British Aerospace/Royal Ordnance and the US firm General Dynamics in tank production), but on the whole the sheer scale of the US defence market relative to that of Europe suggests that interest in such transnational links may be more likely to come from Europe than from the USA in the immediate future. The point, however, is that concentration, with a strong international flavour, is the

Support Group, *Future Relations Between Defence and Civil Science and Technology*, SPSG Review Paper no. 2 (SPSG: London, 1991). The effect of this correction would presumably be to double the proportion of 'true' R&D thought to have spin-off potential, since what would now be removed from account would be final development and testing work, of inherently low spin-off potential.

[16] Among many sources on this theme, see: Walker, W. and Gummett, P., 'Britain and the European armaments market', *International Affairs* (London), vol. 65, no. 3 (summer 1989), pp. 419–42; Taylor, T. and Hayward, K., *The UK Defence Industrial Base: Development and Future Policy Options* (Brassey's: London, 1989); Jackson, J. H., 'Reshaping the defence industry', *Jane's Defence Weekly*, 25 Nov. 1989, pp. 1153–55; Taylor, T., 'Defence industries in international relations', *Review of International Studies*, vol. 16, no. 1 (1990), pp. 59–73; Moravcsik, A., 'The European armaments industry at the cross-roads', *Survival*, vol. 32, no. 1 (1990), pp. 65–85; Anthony, I., Courades Allebeck, A. and Wulf, H., SIPRI, *West European Arms Production: Structural Changes in the New Political Environment*, A SIPRI Research Report, Oct. 1990.

order of the day. This will influence the structure and location of technological capabilities, alter the distribution of national technology bases (and indeed call into question the very concept of 'national' technology bases), and raise questions about the capacity of governments to influence technological development in the face of powerful international corporate interests. This industrial restructuring also raises major questions about the overall regulatory capacity of governments and international bodies (see chapter 3).

Despite the general trend towards internationalization, the grand technological defence project, often international in scope, may be about to disappear (unless it is replaced, as in the context of space technology, by quasi-military projects). Procurement agencies have in recent years risked becoming prisoners of large, increasingly complex development programmes, their budgets becoming vulnerable to recurrent failures of cost control and, in the case of collaborative projects, to disagreements over requirements. The Nimrod, EFA, the United Kingdom Air Defence Ground Environment (UKADGE) and Rafale programmes all offer examples of such difficulties. If procurement budgets are about to decline substantially, the costs and risks of such large projects will become still less supportable, not least because they may squeeze out smaller but no less important initiatives. It seems doubtful, for example, whether the plans for the EFA, proposed in the early 1980s, could succeed if brought forward in the 1990s.

A final issue is whether the opportunity will be taken to redeploy or convert some of the resources currently applied to defence R&D to civil purposes. As mentioned above, in the West European countries and the USA, there is no ready mechanism for the automatic transfer of resources from a reduced defence R&D budget to civil budgets. Traditional technology policy analysis would therefore suggest that much will turn on national policy styles, so that in France, for instance, any cutbacks in defence technology spending will tend to be seen within the framework of a wider concern with a national technology base. There is some evidence that this debate has already begun in France.[17] In the UK, on the other hand, where there has been official concern since the early 1980s over the level of spending on defence R&D, the Government has resisted calls from its scientific advisers to widen the scope of defence R&D,[18] and an experiment to transfer technology from the defence research establishments through the medium of a special company, Defence Technology Enterprises, has been all but closed down. In the USA as well the demotion in April 1990 of the head of the US Defense Advanced Research Projects Agency (DARPA), apparently for showing undue enthusiasm for the idea that DARPA might provide a *de facto* technology policy under an administration that rejects such

[17] Blunden, M. 'European armaments collaboration: what form and what for?', *Science and Public Policy*, vol. 17, no. 3 (1990).

[18] See Advisory Council on Science and Technology (note 14) and the Government's reply in the same document; Science Policy Support Group (note 15).

thinking,[19] suggests that arguments for redeployment of defence resources are as yet received with little interest.

Whether a national policy styles approach is analytically appropriate must also be considered. Gummett and Reppy have suggested that the analysis of responses to the current defence changes can usefully be done through the language of 'policy networks', as developed in the field of international comparative policy studies.[20] The essence of the approach is to conceive of networks of actors (primarily industrial and governmental) organized around specific projects and programmes, and behaving according to a prevailing set of policy norms. The argument then is that the composition of the networks and their norms of behaviour operate as intervening variables through which the impact of changes in the strategic environment, both military and economic, are mediated. Within the networks, government policy influences the behaviour of the defence contractors but, by the same token, the range of feasible policies is constrained by the characteristics of the networks, which may increasingly be international, and by company-inspired changes within them.

Network analysis suggests that not too much attention should be paid to the high-level policy pronouncements of government on the issues discussed in this chapter. Instead it encourages a detailed scrutiny of particular sectors (materials, electronics, telecommunications, and so on) for signs of practical action by governmental and industrial actors to build coalitions that can support specific projects and programmes. In the new climate, it is more fruitful to look for increased activity within these networks by 'defence players'. Signs of such activity are already apparent in calls by leading aerospace companies for major new civil projects. Equally, however, defence players may be expected to seek to influence specific policy areas in a bid to consolidate their position in defence markets. Speaking in January 1990, Mr Battistella, then of the French firm Matra, argued:

> The necessary reconversion of the arms industries is perceived, in a confused manner, by public opinion as an ineluctable reduction in size, while rather it is necessary to prepare for a change of state ['un changement de nature']: the arms limitation treaties contain clauses for quantitative limitation but cannot eliminate qualitative improvement. Every treaty signature has therefore as an immediate consequence a significant growth in the credits for research and development in order to have armaments with a higher performance and to avoid [the possibility] that the adversary could effect technological breakthroughs capable of changing the nature of the strategic dialogue or even of destabilising it.[21]

Down in the networks, it is clear that much is still to be played for!

[19] Boatman, J., 'DARPA director sacked by DoD', *Jane's Defence Weekly*, 5 May 1990, p. 844.

[20] Gummett, P. and Reppy, J.,'Military industrial networks and technical change in the new strategic environment', *Government and Opposition*, vol. 25, no. 3 (1990). This work draws on Wright, M., 'Policy community, policy network and comparative industrial policies', *Political Studies*, vol. 36, no. 4 (1988), pp. 593–612; and Wilk and Wright (note 10).

[21] Cited in Blunden (note 17), p. 7.

Part III
Country analyses

7. The British restructuring experience

Trevor Taylor

I. Introduction

The British defence industrial sector, that is, those companies providing defence ministries with the great range of lethal and non-lethal equipment needed for their armed forces, has not had a stable structure since World War II. In general it has been marked by a trend towards concentration and by governments adopting varying positions on the desirability of government ownership. Since the beginning of the 1980s the biggest single factor affecting the changing structure of the British arms industry has been the procurement policies pursued by the Ministry of Defence (MOD, see section II). The wider defence and economic policies of the British Government have also had an impact (sections III and IV). The behaviour of other governments, especially in Europe, North America and the Middle East, has also had an impact, as has the initiative shown by companies both within the UK and outside (section V), but the core element behind change has usually been the MOD, and occasionally the Cabinet as a whole.

This does not imply that governmental policy has been particularly coherent or successful; indeed, corporate reactions to the official policy environment in which they have found themselves have often exposed the limitations of government declaratory positions and decisions involving resources.

While Government policy represents the biggest single influence on the structure of the British arms industry, it is also important to stress that Government policy itself has been driven by wider developments external to the UK. In fundamental terms, until Soviet General Secretary Mikhail Gorbachev's 1988 UN speech announcing major Soviet unilateral cuts in conventional forces, NATO governments perceived the USSR as presenting a growing conventional threat in terms of the quality and quantity of its equipment. Because it was politically impossible to increase the share of the deterrence burden carried by nuclear forces, Western governments sought to increase their conventional force strength. Arms industries were thus under pressure to generate equipment which was superior in performance and yet affordable. As the fixed R&D and other costs associated with such equipment rose, arms firms needed to act on a larger and larger scale in order to spread large fixed costs over more units produced.

Since 1988 the prospects for conventional arms control have improved massively and the Western sense of threat about former Warsaw Treaty

Organization (WTO) forces has been reduced almost to zero. NATO defence budgets, including British spending, have begun to stagnate or fall. Thus British arms industries, having been asked between 1979 and 1988 to structure themselves so as to be efficient suppliers of technologically advanced equipment to meet a growing threat, have needed since 1989 to structure themselves so that they could survive in a shrinking market.

The formal basis of the argument here is that four lines of Government policy have had an impact on the structure of British arms industries:

1. the emphasis on the need for 'value for money' in defence procurement, which has led to Government enthusiasm for (*a*) collaborative projects, (*b*) competition and (*c*) privatization;
2. the decreasing readiness to spend more money on defence;
3. Government support for British arms exports;
4. Government public indifference to the British defence industrial base.

Responding to these policies and wider developments in the world, British arms manufacturers have reacted by:

1. diversifying to reduce their dependence on defence contracts;
2. making mergers and alliances to remain competitive in specific defence sectors.

In conclusion, the increasing internationalization of the 'British' defence industrial sector is noted, as is the problem of uncertainty which arms producers now face. Uncertainty is perhaps harder to deal with than a declining but predictable market.

II. Value for money

By far the most important factor in terms of its industrial impact was the push announced in the early 1980s for greater 'value for money' to be obtained from the defence equipment budget. The commitment to value for money, which was accompanied by the appointment of an outside businessman, Peter Levene, first as an adviser to the Defence Minister and then as Chief of Defence Procurement, was driven by a blend of three considerations. One was a wish for the Government to be seen not to be wasting the taxpayers' money by paying more than the minimum needed for defence equipment. Another was a feeling that if British firms learned how to compete successfully against each other in the UK, they would have a better chance of competing successfully against foreign firms in export markets. A third and, arguably, the most important factor was a wish to avoid another major defence review. John Nott, Defence Minister before the UK–Falklands/Malvinas War, had carried out such a review soon after Mrs Thatcher came to power in 1979. The 1981 plan for naval cuts in particular caused a major argument in the UK, not least among Conservative members of Parliament. After the UK–Falklands/Malvinas War in 1982, although the defence budget continued to rise, equipment costs tended to rise even more quickly and authoritative assertions were sometimes made that a further review of the UK's defence roles would soon be needed. The 'value for money' emphasis, that is, the holding down of the rate of increase in equipment costs, gave the Government hope that it could avoid making politically painful, structural changes to British defence policy such as abandoning a strategic nuclear deterrent, greatly cutting back on British land and air forces in the Federal Republic of Germany, or reducing the fleet from its strength of about 50 frigates and destroyers down to perhaps 35. One argument against the plan to procure the Trident missile system was that it would divert resources needed in the conventional sector. The political need to avoid a defence review was great when the Soviet threat was growing. Once the Soviet threat sharply diminished after the fall of the Berlin wall, the British MOD came under pressure to demonstrate that it was adapting to the new situation. An 'options for change' exercise was undertaken, and provisional plans to cut the size of British armed forces were announced in the summer of 1990.

Driven by such considerations, the stress on value for money produced three sub-lines of policy which were of consequence for the structure of the British arms industry: collaboration, competition and privatization.

Collaboration

The Government recognized that, for most major systems, the UK could not fund a national development and production project and would have to look for partners to share the burden. Collaborative projects, despite the management problems they presented, were seen by the Government as the way to get access to the capital, technology and markets of others. Thus the UK became a participant, at least for a while, in a series of helicopter projects (the EH-101, NH-90 and the Tonal Light Attack Helicopter), in the Trigat anti-tank missile programme, in the Advanced Short-Range Air-to-Air Missile (ASRAAM) and most notably in the European Fighter Aircraft (EFA). Although Michael Heseltine as Defence Minister in 1982–85 was particularly keen on projects with other European states, the UK was ready to take at face value any expressions of US readiness to collaborate. Hence a number of collaborative projects were begun with the USA, and the UK was an active participant in projects initiated under the so-called Nunn–Quayle Amendment, passed in 1985. In all, collaborative projects accounted for about 15 per cent of British defence equipment spending, according to the Government. Clearly arms control, *détente* and budgetary developments from the beginning of 1989 have contributed to the disruption of many collaborative projects, but British involvement in the spring of 1990 remained extensive, as the list of projects published by the British MOD showed. The UK was a participant in 17 collaborative projects in production or service and in 23 projects in development or earlier study phases. Of the 17 projects, the USA was involved in 6 and, of the 23, it had a part in 13. The UK's main partner in Europe was France (with whom it worked on 21 projects in all), closely followed by Germany (18 projects in all).[1]

Competition

At the centre of the Government's effort to hold down equipment prices was a commitment to get different companies to compete for development and production contracts. The cosy relationship of the MOD to some companies, used to being treated as favoured MOD suppliers of a particular sort of kit, was ended, and the Ministry began to treat its suppliers less as partners and more as servants—and servants who could not supply what was needed at the specified time and price could not expect new orders to be placed. The emphasis on competition also meant that, whenever possible, cost-plus contracts were avoided, and fixed- or firm-price contracts used instead. By the end of the 1980s the emphasis on competition had moved as far as it could and had reached a plateau. In 1988–89 the MOD placed 28 per cent of all contracts by value by competition, 21 per cent using prices related to market forces, 42 per cent using fixed or firm prices set at the beginning of the contract or soon after, and 9 per cent using a cost-plus basis. Six years earlier the equivalent for the last figure was 19 per cent.[2]

Clearly a commitment to competition had implications for industrial structures, although the Government was rather reluctant to face the issues. In some fields it was difficult to have competition as in the 1960s and 1970s a process of government-encouraged industrial consolidation had gone forward in many defence-related industries. In jet engine, electronics and aerospace technology, the UK had its national champions in Rolls Royce, GEC and British Aerospace (BAe). Where it was faced with a British national *de facto* monopoly, the Government took a range of actions. In collaborative projects—in particular the EFA, where the only alternative was for BAe to lead for the UK on the airframe and Rolls Royce on the engine—it encouraged maximum competition for the supply of subsystems. Thus EFA's development has been marked by the establishment of many competing international consortia for various subsystems. Because the UK has a larger and less consolidated defence industrial base than its EFA partners, in many of these consortia there have been competing British companies, whereas the non-British companies in each have been identical, competing only against themselves. In the controversial case of the aircraft's radar, until Siemens at a late stage appeared as a Ferranti partner, the competition was essentially between GEC and Ferranti.

[1] British Ministry of Defence, *Statement on the Defence Estimates 1990* (Her Majesty's Stationery Office: London, 1990), Cmd 1022-I, vol. 1, p. 35.

[2] British Ministry of Defence, *Statement on the Defence Estimates 1990* (Her Majesty's Stationery Office: London, 1990), Cmd 1022-II, vol. 2, table 2.9.

Just as the UK has pressed for competition in collaborative projects, so it has pressed leading contractors on national projects, especially those holding monopolies, to run competitions for the supply of subsystems. Even when a main contractor is the only national choice for a system, for instance as GEC remains for torpedoes, competition can be used to hold down costs in the subcontracting process. The MOD likes main contractors to subcontract 70 per cent of the value in this way. This means that different subsidiaries of a big company, such as BAe, may be called on to compete against outside companies for contracts to be let by other BAe elements. This has implications for corporate organization and the spread of information, since competitions in which companies run must be legally defensible as fair. Competition organized by leading contractors also has meant that foreign bidders can receive consideration, since British private firms may not be committed to use other British firms as suppliers. Some leading British arms companies are understood to be subcontracting on a small scale in the newly industrialized countries of the Pacific rim.

On some occasions, when there has been only one possible British supplier, the Government has been especially ready to consider foreign offers. As a driving force in the efforts of the Independent European Programme Group (IEPG) to open up national defence procurement to all member state firms, the UK sometimes (but not always) treats foreign bids without discrimination. Certainly the UK led the way with the publication of a national *Contracts Bulletin*. Although Vickers finally won the contract for the next British tank with the Challenger 2, the competing Abrams, Leopard 2 and Leclerc tanks were strong contenders. The Vickers management knew that the company could not afford to stumble. Most dramatically, in 1985 the UK abandoned the Airborne Early Warning (AEW) Nimrod project (dominated by GEC) after spending over £900 million (nearly $1.2 billion) and instead bought the Boeing E-3A Airborne Warning and Control System (AWACS) from the USA.

By committing itself in general to competition, the MOD has sent clear signals about it wanted to see in the structure of the British arms industry. It has worried about further concentration in the industry and it successfully opposed the take-over of Plessey by GEC in 1986. When GEC launched a joint bid with Siemens for Plessey in 1988, the UK went along with the proposal with some reluctance and only after some conditions had been met. In general, new firms have been encouraged to become suppliers to the MOD, which has established an office that guides companies on MOD procedures. Foreign companies wishing to enter the British arms market now have access to the help of that office. Thus the number of firms on the MOD's official list of qualified companies has gone up in the last decade and now stands at around 9000.

Although the Government does not acknowledge the fact, many industrialists feel that some companies have been given preferential treatment, in order to bring them into or to keep them in a defence sector. The award to Shorts of the contract for the Tucano trainer aircraft, which helped keep that company in the airframe business, has been cited as an example of such preference.

Emphasis on competition must be considered in the light of the tendency of many defence sectors to be natural monopolies in which the number of potential suppliers shrinks over time. The USA has handled this problem by implicitly or explicitly sharing out contracts to keep companies going. Boeing and General Dynamics both got cruise missile contracts; Pratt and Whitney and GE share combat jet engine work; Hughes and Raytheon each get a viable amount of missile work, and so on. The UK has not practised any such policy except perhaps in the shipbuilding sector and, at least since the late 1980s, has argued instead that the MOD can choose among competing companies from the IEPG states as a whole (and from North America). Thus it does not need to maintain scope for competition in the UK.

The Conservative Governments have preferred to emphasize not that some companies will tend to fall in the arms marketplace, but that under the stimulus of competition British arms firms will become efficient and prosperous. This has left Government policy in a state of incoherence when arms firms have failed or been swallowed up. This was particularly apparent when Westland needed rescuing in 1985–86 and when Ferranti needed help in 1989.

Privatization

The UK has been unwavering in its belief that arms industries (indeed, almost all industries) are best managed from the private sector rather than being in government ownership. This has required it to bring about a radical change in the structure of British defence-oriented firms, that is, to denationalize them. Six defence-oriented companies, each with its own history and character, have been put into the private sector by the Thatcher Governments.

Rolls Royce was taken into state ownership in 1971 after its commitment to the development of the RB-211 jet engine had forced it into bankruptcy. It was sold off in May 1987 through a public share subscription. BAe, the national champion airframe and missile company, was the creation of the Labour Government, which in 1977 nationalized and merged the former BAC and Hawker Siddeley Aircraft companies. The Conservative Government privatized it in 1981, again through public share issues. British Shipbuilders was also set up by the Labour Government in 1977 from a series of companies brought together and nationalized in an effort to slow down the fall of a declining industrial sector. The Conservatives sold off the naval yards separately on the grounds that they might have a commercial future. VSEL bought the mainly submarine-building yard at Barrow and later took over the Cammell Laird yard in Liverpool. The Swan Hunter facilities in Newcastle upon Tyne were bought by the local management, as were the facilities at Vosper Thornycroft on the south coast, which built up its expertise in minesweepers/minehunters and fast patrol vessels. The Yarrow Yard in Glasgow was taken over by GEC.

All these companies needed debt to be 'adjusted' before they could be sold, but those that could be sold by public issue of shares looked the most sound in future commercial terms. A more difficult case was that of Royal Ordnance (RO), which was traditionally a state arsenal supplying British armed forces rather than a commercial organization. From 1983 this was made into a more autonomous body and in 1985, to prepare it for private ownership, it was turned into a state-owned company. It became clear that it would be impossible to float it successfully on the stock market and so companies were invited to bid for all the shares. In a much disputed development because of the value of the land which RO owned, BAe submitted the best bid of £190 million ($311 million) and took over RO in April 1987; RO now operates as a separate entity within BAe. In a similar and parallel development, BAe also submitted the best bid for Rover cars (including Land Rover) in 1988. As for naval repair and maintenance, having announced the closure of one Royal Dockyard at Chatham, the Government kept ownership of the other two (at Devonport in the south-west and Rosyth in Scotland) but sought private companies to manage the dockyards' operations for a fixed period. Contracts were awarded to two engineering companies, one of which, the US-based firm of Brown and Root, led the group managing the Devonport yard.

Finally, the Northern Ireland aerospace company Shorts, with its capabilities in anti-aircraft missiles and small civil transports, was sold in 1989 to the Canadian Bombardier company, in competition with GEC and Fokker.

This extensive and multifaceted privatization programme clearly raised among other issues the question of foreign ownership. To what extent was the Government ready to see British arms producers come under foreign control? The initial answer was not much, at least with regard to the larger firms, since with BAe and Rolls Royce a limit of 15 per cent was put on foreign-owned shareholdings. In 1989, after the companies protested that they would be worth more if they could sell more shares on the world market, this limit was raised to 29 per cent. Also, after the 1985–86 Westland affair,[3] the Government was not keen to sell Land Rover to foreign buyers on political grounds. On the other hand, as noted above a US firm was permitted to run Devonport, and there was no special effort to find a British buyer for Shorts. Siemens, Thomson-CSF and Hughes are among the foreign companies which have been allowed to buy arms companies in the UK.

Thus, in a number of ways, the emphasis on value for money in defence procurement, using privately owned firms which would compete and collaborate to win and execute defence contracts, has had substantial implications for industrial structures. One such implication was that British firms were encouraged to adopt an international orientation, partly by the nature of collaborative projects which required a trusting close partnership with foreign firms if they were to be successful. However, competition and the need to hold down costs encouraged British main contractors to find foreign partners who could provide less expensive technology and subsystems. The competition for

[3] See chapter 2, pp. 26–27; see also Linklater, M. and Leigh, D., *Not Without Honour: The Inside Story of the Westland Scandal* (Sphere Books: London, 1986).

an anti-armour weapon for the Tornado aircraft showed what could happen when a range of British firms placed bids together with foreign partners. The competitors included ML Aviation of the UK, bidding in collaboration with Rheinmetall; Hunting, bidding with Diehl and Honeywell; BAe, with MBB and Hughes as partners; and GEC, working with Rockwell.[4]

III. Economic developments and overall defence policy

The structure of British defence-oriented industries has been affected, not just by procurement policy, but also by the wider consequences of defence spending levels in conjunction with broader official economic developments and policies. Because the UK has since World War II tended to spend more than 5 per cent of its gross national product (i.e., more than the EC average) on defence, and because British overall manufacturing growth since World War II has been less impressive than that of most West European states, the defence share of British manufacturing has consistently been significant. However, at least during the first seven years of the Thatcher Governments, there were two parallel processes under way. One was the decline in British manufacturing caused by the international depression, and by the British Government's readiness to accept multiple bankruptcies and high unemployment as the price of getting inflation down and weakening the power of the British trades unions. The other was that the UK endorsed the need of NATO to spend more on conventional defence so that conventional forces could take a bigger share of the deterrence burden in the light of increasing WTO capability (this may be a rather grand rationalization covering up the basic fact that British equipment spending went up to pay for the purchase of the Tornado and Trident systems). Since 1986 British equipment spending has tended to stagnate and even decline in real terms, while manufacturing output in general has recovered. The consequence is that today arms production (which is not easily defined) accounts for 8–10 per cent of British manufacturing output, with exports being taken into account. This is a substantial proportion.

The Government has shown little direct concern with this figure so far, but it has concluded that defence takes too great a share of the British research and development (R&D) effort.[5] In broad brush terms the official statistics shows that about half of British official R&D spending goes to defence, while privately funded British R&D runs at about the same level as the total Government effort. In round figures, the Government spends just over £2 billion ($3.4 billion) a year on defence R&D, and similar amounts are spent both by industry and government on civil R&D. The Government has concluded that too much official R&D is going to defence and has been seeking successfully for the past couple of years to decrease that share. Its hope is to stimulate

[4] Taylor, T., 'Defence industries in international relations', *Review of International Studies*, vol. 16, no. 1 (Jan. 1990), p. 64.

[5] The Cabinet Office, *R&D: Annual Review of Government-Funded Research and Development* (Her Majesty's Stationery Office: London, 1989).

technological innovation in the civil sphere, but some, led by the Society of British Aerospace companies and in particular by BAe Deputy Chief Executive Ivan Yates, argue that the Government has misread the situation. They argue that the basic problem is that British industry will not fund enough civil R&D, and that much of the Defence Ministry's R&D budget is not spent on innovative work that meets the Frascati definition of true R&D, but on routine testing and development work.[6]

Whatever the merits of this particular case, Government defence R&D spending has gone down, and the Government's half-dozen specialist research centres are being merged into a single organization which will have to go out and win research contracts. Part of these establishments' past job has been to provide the Government's 'expert customer' capability, by which equipment proposals can be evaluated. Private companies, including systems houses, already get some such work and in future they will be able to compete for more.

Clearly there are questions about what level future defence spending will reach. Although the British Government at first reacted prudently to improvements in East–West relations, it recognized that a peace dividend should be forthcoming. It refused to use the Iraqi invasion of Kuwait as a reason for sustaining defence spending and announced in 1990 that the defence budget would fall by 6 per cent in real terms by 1994. In July 1990 it announced the outline of a new structure for the British armed forced, some of the details of which were given in the 1991 Defence White Paper. Army forces were to be reduced from 156 000 to 116 000 men. Overall, planned service personnel cuts were greater than envisaged overall spending reductions (by 6 per cent in real terms in 1993–94 compared with 1990–91). The implication was that the procurement budget would take a less than proportional share of the total budget reduction.[7] A Labour Government, it can be judged, would also not be keen to cut defence spending to the point where industry would be substantially disrupted.[8] In broad terms, military expenditure can be expected to decline slowly in the next few years, assuming that East–West relations continue to improve also. This is a central feature of governmental behaviour to which companies will have to adjust and in some cases have begun to do.

IV. Government support for arms exports

While changes in government defence spending patterns have benefited and then put stress on British arms industries, the Government has been consistent in providing political and organizational support for arms sales overseas. When she felt it could be of benefit (and she did not always judge correctly), even

[6] See PA Consulting Group, *The Significance of R&D Investment to the UK Economy* (PACG: London, 1988).

[7] British Ministry of Defence, *Statement on the Defence Estimates: Britain's Defence for the 1990s* (Her Majesty's Stationery Office: London, 1991), Cmds 1559-I and 1559-2, vols 1 and 2, especially vol. 1, paragraphs 404–24.

[8] See 'Planning for peace puts priority on jobs', *The Independent*, 14 May 1990.

former Prime Minister Thatcher sought to help a company win a contract. More consistent and day-to-day help is forthcoming from the Defence Export Services Organisation (DESO), which although a government body is always headed by a seconded industrialist.[9] The DESO cannot only help companies with advice and contracts, it can also contribute to a Government decision to purchase a package of British (and if appropriate non-British) products to meet the requirements of its overall defence policy. This was the case with the al Yamamah sale to Saudi Arabia[10] and with the major sale to Malaysia that is only slowly being hardened up and clarified.[11] The DESO not only stations staff in the UK while sending them overseas periodically, it also will post staff to customer countries to help secure a sale.

Overall, the support which the Government gives to British arms exporting efforts has helped keep the level of British exports reasonably stable in an era of declining demand. British defence orders in 1986 totalled £6 billion ($8.8 billion) and in 1987 and 1988 the level of contracts was around £3.5 billion ($5.4 billion) per year, enough to keep the UK in third place behind the USA and the USSR.[12] Another factor here has been that the competitive fixed-price environment which the MOD has established at home has contributed to making British arms producers into leaner, fitter and more commercially-minded enterprises that market effectively overseas.

V. British defence industrial base policy

The picture so far painted is one of British defence industrial structures which are private and accustomed to competition. They account for a sizeable share of British manufacturing. However, a significant change regarding Government policy is that the Government has lost nearly all commitment to maintaining a national defence industrial base for its own sake. There is no policy of maintaining a national capability in any area other than one where the UK cannot buy appropriate equipment from overseas. An MOD classified survey in the 1980s revealed only four such areas and their number is not thought to have increased since then. Any sense that the UK should maintain a capacity to increase production of heavy consumption items such as ammunition in the case of war has been given up, since maintaining such a capacity costs money. Outside of the foreign share limitations on BAe and Rolls Royce, it has rarely been thought important in principle whether arms manufacturing facilities in the UK are owned by foreigners. This lack of formal concern with the

[9] For a largely positive report on the performance of DESO, see the National Audit Office, *Ministry of Defence Support for Defence Exports* (Her Majesty's Stationery Office: London, Apr. 1989).

[10] Outlined in British Ministry of Defence, *Statement on the Defence Estimates 1989* (Her Majesty's Stationery Office: London, 1990), Cmd 675-I, vol. 1, paragraph 416.

[11] See 'Malaysia gives commitment on £1 billion defence package', *Financial Times*, 14 June 1991. See also chapter 2, p. 22 in this volume.

[12] National Audit Office (note 7), pp. 1, 6–11.

capabilities and control of the British defence industrial base is a specific feature of government policy under Mrs Thatcher.[13]

In practice, foreign ownership has become a feature of some prominent, if second rank, arms contractors. However, the Government has seemed ill at ease with sudden or unannounced take-overs and mergers. This was most apparent and important in the Government's handling of the 1985–86 Westland crisis, in which there was disagreement in the Cabinet as to whether Westland should become dependent on United Technologies. However, governmental uncertainty was also an element when GEC in 1989 joined with Siemens to take-over and share out Plessey, and in 1989–90 when Thomson-CSF emerged as one of the companies which was interested in taking over all of the troubled Ferranti company. In each case the Government took a pragmatic view, indicating privately that there were parts of Plessey and Ferranti which they did not mind going into foreign ownership but parts that they did. London did not want all of Ferranti to be taken over by a foreign, state-owned firm. Saving difficulties for the Government, GEC took over Ferranti's airborne radar business, leaving Thomson-CSF to buy Ferranti's sonar business (and Philips' British MEL simulation business) in 1990. In principle, the Government does not object to foreign ownership of British arms firms, but in practice does not always like the prospect.

VI. Corporate reactions

What kind of industrial structures are emerging in the light of the above policy features of the British Government? How are British companies reacting to the changes of the past decade which have required them to be competitive producers of high-technology arms, most recently in a declining and uncertain market?

Diversification

A central factor here is that although much of British industry is defence-related, the UK has few specialist arms firms outside shipbuilding. Most of the leading companies earn at least half their income from the civil sector, and Government procurement changes have also tended to make the defence sector less profitable. On the other hand, many firms have fairly autonomous sub-divisions that are mainly oriented towards military production.

It is not easy to generalize about how firms are responding to the new British and wider political environment. Companies which are small in military production may seek to hang on, as GKN is doing on the basis that the demand for light armoured vehicles will be sustained. GKN makes the Warrior infantry fighting vehicle, but arms production accounts for only about 10 per cent of its

13 For a discussion of this issue, see Taylor, T. and Howard, K., *The British Defence Industrial Base: Issues and Options* (Brassey's: London, 1989).

turnover. Thorn EMI and Racal, on the other hand, sought unsuccessfully to sell their defence sections because there were other business sectors on which they wanted to concentrate (entertainment and communications, respectively), and because they felt they would lack the technological resources to remain competitive in military electronics.

Predictably, companies are looking to become less arms-oriented through mergers and acquisitions rather than by suddenly introducing civil production into arms factories, a process recognized as difficult. However, BAe and GEC, among others, are encouraging their divisions to look for civil applications of their military technology, and BAe has established a specified Enterprises division whose role in part encompasses such an approach. In another instance, GEC Underwater Systems is looking for civil applications of its technology, such as exploring the sea-bed for natural resources. In aerospace sections of defence, some industrial adjustment is aided by the current boom in the civil aircraft sector. Thus Westland is offsetting a lack of helicopter orders in part by moving more extensively with the civil airframe components business.[14] This is a significant source of work also for Shorts in Belfast.

On the shipping side, British naval yards have been seeking civil work for some time, often with an eye to specialist high-technology ships. In a purely British context the easing of East–West tensions may in the immediate term help shipbuilding, as financial resources earmarked for the British Army are released and perhaps made available to the Royal Navy. Swan Hunter was awarded a contract to build three Type 23 frigates at the end of 1989; the Yarrow Yard still has British frigate orders to fill; and VSEL at Barrow is building Trident submarines. Vosper Thornycroft has export orders to keep it going, but Cammell Laird looks as though it may soon be in real difficulty.

Like most of the UK's arms industry, the shipbuilding sector is seeking to increase exports, but in recognition of the capabilities and preferences of customers, many British shipyards are offering to sell design and management services rather than shipbuilding. On balance, naval shipbuilding remains highly problematic, and VSEL has responded by diversifying, not out of military production but into armoured vehicles. It has developed the AS-90 self-propelled mortar (after the failure of the collaborative SP-70) and has found a customer in the British Army.

Thus there is no uniform pattern of response in terms of diversification or in terms of overall attitudes towards the arms industry. However, there is also a good deal of uncertainty and worry in industry about how NATO defence ministries are likely to behave in the future. Will they opt for equipment which is highly compatible with defensive operations even if it is expensive, or will they shun big-ticket items in general? Will governments make procurement decisions at all, or will they tend to put off making up their minds until political circumstances clarify?

[14] Helicopter sales fell 21% in 1989 and accounted for just two-thirds of Westland's turnover. This proportion is expected by the company to decline further. See the Westland 1989 results report in *Financial Times*, 30 May 1990.

Mergers and alliances

As British (and foreign) defence-oriented firms have looked at the changing political and technological environment of defence, they have often felt the need both to become larger actors and to have overseas links.

Sometimes this has had uncomfortable consequences for individual firms in the UK. The pattern of change to date has already brought concentration in the British military electronics industry as Plessey, against the will of its executive, has been successfully taken over by GEC and Siemens. Ferranti's troubles with its ISC subsidiary forced it to sell many of its military production interests to GEC, with the consequence that the smaller Ferranti is much less dependent on military production. GEC will rationalize and concentrate its airborne radar capability, probably building on the Ferranti base in Scotland, and the UK will have a single source of expertise in this field rather than two.

Governmental actions in trying to open up IEPG and NATO arms markets to international competition and in encouraging collaboration have given a lead. The British national champion companies, like their peers in Europe, see that their future is bleak if they rely only on the support of their national government and wait for collaborative projects to come along. In addition to showing initiative on major collaborative exercises such as the EFA, British firms have not been reticent in seeking more permanent, non-project specific, international links, oriented in three directions: towards North America, Western Europe and the Third World.

Matched perhaps only by Thomson-CSF, several British firms have been ready to invest in military production facilities in North America, so that GEC, for instance, now has 5000 employees in North America. In the civil and military aerospace field, Smiths, Lucas and Dowty are important subcontractors with significant US holdings. British firms are encouraged to invest in the USA both by the size of the US arms market and by the recognition that it is normally necessary to arrange US production if a piece of equipment is to be sold to the US Department of Defense.

In Western Europe, although there has been some British investment in military production by companies such as Astra (buying, in a disastrous experiment, the Belgian PRB company[15]), United Scientific Holdings and by the bigger companies (GEC has operations in Italy and BAe owns a German electronics firm), the joint co-operative venture between big firms is more striking. Links between British and French industry are growing, in particular in the shape of the GEC–Matra joint space company; the cross shareholding arrangements being developed by GEC, Matra and Daimler Benz; and the BAe link with Thomson-CSF to establish the Eurodynamics missile company. Such links of course are interlaced with *ad hoc* collaborative programmes such as the Trigat, the EFA and the Family of Anti-air Missile Systems (FAMS). Mergers and alliances are also related to the needs of the Single European Market in

[15] See 'Astra tentacles spread', *The Independent*, 16 July 1991.

civil goods: Siemens interest in Plessey was a more a function of Plessey's telecommunications business rather than of its defence activities; yet the latter were unavoidably involved.

BAe's chairman, Sir Roland Smith, has said he believes that within a few years European arms industry will be dominated by three or four large groups which will have a strong transnational dimension.[16] As leader of the UK's biggest arms company, he is in a reasonable position to make his prediction come true. In diversification terms, a valid generalization is that defence-oriented firms are looking to diversify not just away from defence products, but also away from dependence on British Government orders. For that, an overseas presence is valuable.

Increasingly this is also being recognized by corporate leaders as valid for export markets beyond the EC. GEC has sold tactical radios to Turkey and will be making them there. It hopes to be able to sell some of its Turkish production into Middle East markets. BAe has set up a joint venture and is looking at arms production in Singapore, and GEC's extant manufacturing arrangements in Malaysia have a defence dimension, albeit minor. BAe has realized that sub-contracting from Far East companies can help keep down costs of the final product and help win export orders later. Whether it is Vickers selling turrets for Brazilian armoured vehicles, or Rolls Royce setting up a joint company in Spain to help Spanish industry reach its 13 per cent share of the EJ200 engine for EFA, British companies are showing that they are ready to establish such international links as are necessary to win orders.

As British companies look overseas, so are foreign companies looking to the UK. While US firms have generally been reluctant to invest in European arms firms after the Westland affair, an exception has been the Hughes take-over of Rediffusion. Thomson-CSF has led Continental European firms acquiring British businesses, and its acquisition of the Link-Miles company means that the simulation business in the UK could well be dominated by foreign firms. Thomson-CSF has also bought Ferranti Sonar, while in the software/systems design field Sema-Cap is a prominent Franco-British company. Clearly the process of European integration and the implementation of the Single European Act is bringing foreign investment to the UK which has an implicit, secondary and even hidden defence dimension. For instance, take-overs have now left the British ball-bearings industry in foreign hands.

Rationalization

By the middle of 1991 some defence-oriented companies were seeking also to survive by rationalizing their production on fewer sites and by cutting employment levels. As an example, in December 1990 BAe decided to close factories in Kingston, near London, and in Preston, Lancashire, involving a total loss of 5000 jobs. It announced a further 4700 job cuts (in missiles and civil aerospace)

[16] Cook, N., 'European industry must "put business first"', *Jane's Defence Weekly*, 14 Oct. 1989.

in March 1991. Soon after the submarine manufacturer VSEL announced 5 500 redundancies in Barrow.[17] Rolls Royce job losses at its largely military business in Bristol could reach 1200 in 1991.[18] Lay-offs became increasingly common, for example with aerospace manufacturer Dowty introducing 13 000 job cuts, many involving its defence sector, in April 1991.[19] While some of these changes reflected a drive for cost-cutting and greater efficiency (at the subcontractor level it was not always possible to distinguish defence-related job losses from those associated with the British economic recession as a whole), they implied that defence manufacturing capacity to a considerable extent was being abandoned. This was clearly the case with companies, such as Fraser-Nash in engineering/aerospace, which went out of business.[20]

VII. Concluding points

No specific conclusion can be drawn from this survey of how the policies and activities of the British Government affect the British arms industry and of how the industry is responding. One theme is clearly that the national character of the British arms industry is being eroded and that multinationalism in various forms is ever more prominent.

Three points emerge, implicitly as much as explicitly. The first is that British companies are not yet panicking to get out of military production, not least because the British defence budget and export markets have not fallen sufficiently to induce panic. Companies are increasingly considering diversification, both into commercially more viable military production and into civil goods, as well as at ways to strengthen their positions through international links. At the same time, the restructuring that has taken place is more the result of governments' greater readiness to see more open competition in European arms production (and by growing resource demands for high technology) than of shrinking arms markets.

Second, the term 'conversion' has not been mentioned. Diversification and adjustment have been preferred as terminology when the transformation of defence into civil business has been discussed. This is partly because conversion only occupies a modest spot on the British political agenda, but also because there is a danger that using the word 'conversion' will lead to over-simplification. A specific comment to be made here about conversion is that initially it may well be necessary to convert not factories but R&D resources. Factories need to make (on a competitive basis) developed products and, while something can be done through the acquisition of licences, in the longer term firms need goods developed in-house if they are to be commercially successful.

[17] See 'BAe may soon shed up to 3, 000 more jobs', *Financial Times*, 21 Mar. 1991; 'British Aerospace to cut 4,700 jobs', *Financial Times*, 22 Mar. 1991.

[18] 'Rolls Royce jobs cuts rise', *Financial Times*, 29 Apr. 1991; 'Rolls Royce doubles jobs cuts', *Financial Times*, 9 May 1991.

[19] 'Dowty jobs to go', *Jane's Defence Weekly*, 20 Apr. 1991 p. 668.

[20] 'Fraser-Nash calls in the receiver', *Jane's Defence Weekly*, 8 Dec. 1990, p. 1130.

Having acquired a civil R&D capability, a firm also needs a civil marketing organization which can work to sell a new product. Retooling a factory to allow a shift from military to a civil production ought to be a near-final stage. For these and other reasons, conversion is thought by many British industrialists to be extremely difficult. In many cases it might be more effective to close a plant down and to start again, hopefully in an economic climate of expansion and low interest rates. None the less it would appear logical that companies which have had to compete vigorously in the defence sphere, as have many British firms, will find it easier to adapt to competition in the civil area.

Third, so far the main problem which arms control and *détente* have presented for industry has concerned the uncertainty which these issues raise. It is some years since John Kenneth Galbraith argued that high-technology business needs to be planned and hence to have a predictable future.[21] In arms production this is particularly valid, since it can take a decade or more for a system even to be developed. Defence ministries today cannot be confident about how they will be acting in the future, and this inevitably makes things difficult for defence-oriented companies anywhere.

In the UK a radical forecast would be that companies implicitly rather than explicitly will turn away from defence, given that it does not appear to offer many chances for growth and that the precise nature of spending is so hard to predict. Companies may still bid for MOD contracts, but may not make their best technologists and managers available for defence work as future turnover and profit are found elsewhere. Already some British firms, including BAe, are promoting an internal culture that seeks to remove the 'defence' label. Instead they are 'electronics', 'engineering' or 'aerospace' manufacturers, serving customers and seeking maximum profit on capital invested. The MOD is on its way to become one customer out of many, important but rather unreliable, and less intimate than others.

How would the MOD react in the long term should its standing with industry continue to decline? Will it pay to get more attention by offering security of business and higher profit margins? In a largely disarmed world, will the armed forces once again be wanting their own sources of supply, their own arsenals? Should defence spending in the UK and elsewhere continue to decline, the consequence will not simply be that the defence industrial sector shrinks. It will also change in character. Will specialist arms firms or high-technology firms with defence interests take the lead? Will national or predominantly multinational arms firms dominate the market? Will arms firms be privately owned or nationalized concerns? Will they have an intimate or a hands-off relationship with their customers? Will governments present defence as a high-risk, potentially high-profit business, or as a secure, low-profit field? It is not yet possible to forecast with any confidence just how the British defence industrial sector will change, but perhaps the dimensions of change at least can be identified.

[21] Galbraith, J. K., *The New Industrial State* (Penguin: London, 1969).

8. Reorientation of French companies

*Claude Serfati**

I. Introduction

This chapter deals with the consequences for the French arms industry of the dramatic changes currently taking place in the world arms markets. The desire, often expressed by the French Government, to continue to develop the national arms industry is confronted by the reality of a decline in exports. This situation will inevitably lead to increased co-operation between French and other European arms producers. For the French firms this marks a turnabout in their arms development and production strategy.

II. Towards European co-operation

Among European countries, France is the most affected by the profound developments that have taken place in the international arms market in the 1980s. This is in part due to the role that the French arms industry traditionally has played as an instrument of Government industrial policy, in part to the industry's export-oriented strategy, focused on the Third World, and in part to the high concentration on French suppliers of components. These factors are briefly outlined below.

Developments in the international arms market

Three developments in the international arms market are of particular importance to France. The first is the reshaping of US military expenditure and production patterns that occurred in the 1980s. Faced with a decline in economic competitiveness, the Reagan Administration saw in the increase of military research and development (R&D) expenditures and the launching of major system programmes a way to restore US leadership in advanced technology.[1] For arms producers as well as for other governments, the US focus on costly high technology has meant that the level of military spending

[1] Haudeville, B., 'Leadership technologique et R-D militaire: les politiques des années 80', ed. F. Chesnais, *Compétitivité et dépenses militaires* (Economica: Paris, 1990), pp. 1–44.

* This paper is part of a wider research project at the Laboratoire de recerche en économie appliquée–Centre d'etude et de recerche sur l'entreprise multinationale (LAREA–CEREM), conducted jointly with François Chesnais. I am grateful for his comments on earlier drafts.

necessary to remain competitive in this sector of the arms market was raised. The likely future reduction in US military procurement spending will likewise impact heavily on European arms production and procurement world-wide, in two ways: (*a*) European firms might increasingly be closed out from the all-important US market;[2] and (*b*) European firms will be faced by more acute pressure as US firms attempt to penetrate the European market.

The second development is the decline in arms sales to the Third World, as a result of the persistent and deepening economic crisis experienced by major Third World buyers. This is of particular concern to the French arms industry, as the main part of its exports has been oriented towards Third World markets.

Third, the spread of local, low-intensity conflicts has provided new arms exporters, mainly Third World producers, with a growing market for cheap, less sophisticated weapon systems. The Third World thus accounts for changes on the export as well as the import side.

These developments outside of Europe were already under way when the processes of political change in Eastern Europe culminated in 1989–90. The latter will likely further modify the conditions and orientation of the European arms market. However, the impetus for disarmament and for a reorientation of NATO military planning following these changes notwithstanding, the political and economic importance of maintaining stable defence structures seems to discourage governments in leading industrial nations from allowing military expenditures to fall as much as changes in security perceptions would warrant.

The French arms industry as an instrument of Government industrial policy

The development of a national arms industry, intimately linked to a strong military tradition, served as an impetus to the growth of French industrial capitalism in the 18th and 19th centuries. More importantly, military influence crossed over into political life and helped establish, in the early 19th century, the basis for a distinct Government science and technology policy. Several aspects of this policy are still pursued.[3]

Since the 1960s the arms industry (including the large nuclear sector and later the space technology sector) has been at the core of Government industrial and technological policy. It has become a priority sector to which others have had to cede; sectors which could not adapt to the new demands quickly faded.[4] Imposing military budgets, in addition to generous tax allowances and export credits, have provided the national arms industry with a strong economic base, allowing a dozen large firms to prosper. As a result, what may be called the

[2] Serfati, C., 'Stratégies et contraintes d'un groupe français à spécialisation militaire', in Chesnais (note 1), pp. 217–45.

[3] Serfati, C., 'L'économie française et le fardeau des dépenses militaires', *Temps Modernes*, no. 524 (Mar. 1990), pp. 122–67.

[4] See Chesnais, F. and Serfati, C., 'L'industrie militaire, une locomotive du développement industriel?' in Chesnais (note 1), pp. 169–216.

oligopoly of the French arms industry has grown around the Délégation Générale pour l'Armement (DGA), the government agency responsible for military R&D and production. This oligopoly, structured by tightly interwoven industrial and financial links, as well as by personal relationships between government and industrial groups, has remained relatively intact throughout the 1980s. Around this oligopoly a small number of middle-sized firms, with annual revenues of 1–4 billion French francs (roughly $160–660 million), have been able to survive thanks to their technical excellence in certain niches.

The economic role of military production is coupled with the political role that the Army has played in French colonial expansion and in defence of French colonial rule and subsequent post-colonial dominance.[5] This tradition has proven notably resilient. Despite dramatic changes in the arms market in the 1980s, the emergence of a new social and political order in Eastern Europe and in the USSR, and progress in international arms reduction talks, French officials obstinately continue to affirm the need to maintain military expenditure levels, allowing only marginal cuts. They argue for continued funding of current R&D programmes, envisaging new generations of existing systems.

Despite the generous state credits provided in the recent past, if the French arms industry is to continue with the programmed projects, with an average development and production time of 15–20 years for each major system, it will inevitably be forced to modify its strategy.

Reorientation of sales and production

In simple terms, this means that the French arms industry must reorient its sales and production effort to the European market. Three tasks top the agenda:

1. The French arms industry needs to gear its efforts from shrinking Third World markets, which have hitherto been the main outlet for French arms exports, to the cash-rich markets of the European countries.

2. French firms need to increase co-operation with US or European firms to share the high costs associated with the development of new and sophisticated arms systems. Given that US firms tend to dominate in US–French partnerships, French firms, in pursuit of a more favourable environment, will likely prefer co-operation with European firms.[6] However, even such European co-operation will mean that France will lose the leadership role to which it is accustomed.

3. Firms must become sufficiently large to rival British, German and US competitors. Although France remains the world's third largest supplier of

[5] See Chesnais, F. and Serfati, C., *L'industrie française d'armement* (Circa Nathan: Paris, 1991).

[6] Serfati, C., 'La coopération européenne dans l'industrie d'armement: chemin de croix ou nécessité?', ed. A. Alcouffe, *L'Europe industrielle horizon 93* (Documentation française: Paris: 1991), vol. 2, pp. 81–100.

conventional weapons in national terms,[7] on a company level its 'national champions' are handicapped by insufficient size.

III. Inter-firm co-operation: some methodological observations

Since the mid-1970s, inter-firm co-operation agreements have multiplied in all high-technology industries. Classifications of these agreements are found in a rich body of data base analyses.[8] By their growth and by the large size of the companies involved, these agreements are increasingly seen as complementary to the consolidation of a company's activities, permitting it to cope with the modifications and uncertainties of the global economy. The main factors behind the multiplication of such agreements can briefly be summarized as follows:[9]

1. Developments in science and technology: cross-fertilization between scientific disciplines, the managing of systemic features of new technologies and the emergence of new technological paradigms imply increased R&D costs.
2. The increasing importance of science and technology as barriers to entry into industrial competition: inter-firm technical agreements permit large firms to control the new technological trajectories.[10]
3. Inter-firm agreements allow large firms to gain control over new technologies developed by public research institutes and medium-sized firms.
4. Agreements among large firms provide a formal framework for the interlinked decision processes typical of concentrated industries in which oligopolies prevail.

Some of the factors explaining the growth of co-operation agreements in general have specific application to the arms industry. However, care should be taken not to apply general classifications to arms co-operation agreements. The arms industry differs from civil consumer industries in that it to a greater extent relies on government participation to realize co-operative agreements. Moreover, the long life of co-operative arms projects confers a certain stability to the agreements under the terms of which they are launched. The structure of such agreements implies a precise definition of the roles of each co-contracting company, both in terms of financial shares and of the volume of purchases per country. Production is usually divided into several stages, with each firm

[7] See Anthony, I., Courades Allebeck, A., Hagmeyer-Gaverus, G., Miggiano, P. and Wulf, H., 'The trade in major conventional weapons', SIPRI, *SIPRI Yearbook 1991: World Armaments and Disarmament* (Oxford University Press: Oxford, 1991), p. 215.

[8] The data bases elaborated by the Laboratoire de recherche en économie appliquée–Centre d'etude et recherche sur l'entreprise multinationale (LAREA–CEREM), Université de Paris X (Nanterre campus), are mainly focused on inter-firm agreements between European groups in the advanced technology sector.

[9] This summary is based on the research done by François Chesnais for the Organization for Economic Co-operation and Development (OECD), presented in 'Les accords de coopération technique entre firmes indépendantes', *STI Revue* (OECD), vol. 1, no. 4 (Dec. 1988), pp. 55–132.

[10] The notion of technological trajectories expresses in analytical terms the basic cumulative and evolutionary features which mark the developments and changes experienced by technologies as they are diffused and employed in production and services.

assigned to one task while one company oversees the entire project. Finally, the interaction between the political and economic forces (i.e., the balance between government and industry influence) is such that the agreements concluded have established either the practically parallel co-production of arms or the creation of new corporate entities.

Using the methodological approach described above, co-operation agreements related to the conception and production of arms are here referred to as 'institutionalized co-operation agreements' to underline the decisive role played by governments in effecting and implementing such accords, the stability resulting from their relative longevity, and the rigid organization of the co-operative effort.

Two main types of co-operation may be distinguished, with firms joined either in a senior partnership or in a consortium. In senior partnership arrangements, one of the participating firms assumes a leadership role, and the executive management of the venture as a whole becomes subject to the state controls that apply to that partner. Firms of other co-contracting countries are responsible for the manufacture of specified parts and components. When two countries and/or firms enter into such an arrangement on an equal basis, the senior partnership can be shared on equal terms. Examples of such joint senior partnerships include the Franco-British Jaguar aircraft programme, with Dassault as senior partner in airframe production and Rolls Royce–Turbomeca as senior partner in engine production; and the Franco-British Lynx/Gazelle joint helicopter programme, with Aérospatiale as senior partner for the Gazelle and Westland as senior partner for the Lynx.

The second type of co-operative arrangement, the consortium, is often chosen as a means for managing design and development projects. In this, responsibility is shared between the co-contracting countries, each represented by manufacturing firms that share in the production of the finished design. Such consortia are managed in a collegial way. Few have enjoyed as much success as the Franco-German Euromissile venture, owned equally by MBB and Aérospatiale. Founded in 1972, Euromissile has successfully completed three major projects: two anti-tank missiles (Milan and HOT) and one ground-to-air missile (Roland). The first two were commercial successes. In the case of the Roland, the US purchaser, the US Department of Defense, went to lengths to have the missile redesigned (at a cost corresponding to 2.5 times the amount Euromissile had originally spent on design), and then to sell the new version to its NATO allies. As a consequence, Euromissile's sales and profits on the Roland have remained modest. Another example is offered by the two consortia set up to manage the Tornado fighter aircraft programme. The first, Panavia GmbH (airframe), is jointly owned by British Aerospace, MBB and Aeritalia; the second, Turbo-Union (engine), is jointly owned by Rolls Royce, MTU and FIAT. The breakdown of ownership shares among the companies in each consortium is 42.5, 42.5 and 15 per cent, respectively.

These examples notwithstanding, European co-operation in arms production has hitherto yielded few successes. Not only has the number of co-operative programmes been limited, but just as many projects have been aborted as have actually been launched. More importantly, far from being broad-based European efforts, most of these projects are the fruits of bilateral arrangements. Case studies demonstrate that along with the benefits gained from such co-operation (economies of scale, learning to work in teams), problems—including cost increases, delays and weak sales—marked every stage, in part due to the fact that jointly produced systems have always to compete with similar, nationally produced ones.[11] Such was the case with the Franco-British Jaguar aircraft, which met tough competition from the all-French Mirage. The sale of the latter was given preference by Dassault, the prime contractor, despite that firm's major interests in Jaguar. Instances such as these indicate that European co-operation ventures are the fruit of such fragile compromises of diverging national interests (strategic, industrial and technological) that each case needs to be considered individually.[12]

IV. French interests in European arms industry co-operation

The interest demonstrated both by the Government and by industry in greater European co-operation marks a turnabout of the strategic policy that has guided French arms production and trade since the 1960s. Institutionalized co-operation agreements are now seen as offering a number of advantages both to the public and to the private interests:

1. They provide hope for French producers to enter the new markets (mainly in Europe) needed to compensate for shrinking exports to the Middle East.
2. They offer a competitive advantage that provides a barrier against new entrants into existing markets, as well as push less competitive companies either out of the field or into the arms of larger firms as junior partners. French firms will be able to use this strategy against each other, as in the defence electronics sector, as well as against other European competitors.
3. They seem to offer bright export prospects and economies of scale. This may prove to be an illusion, however, considering the record of past and present European co-operation efforts, which in each case has been characterized by astronomical overcosts.

The French Government has been very active in promoting European co-operation. Following the proposal by President François Mitterrand in 1985,

[11] See RAND Corp., *A Review of European Arms Collaboration and Prospects for its Expansion under the Independent European Programme Group*, RAND Note no. 2638–ACQ (RAND Corp.: Santa Monica, Calif., July 1987); Hartley, K., *Arms Co-operation: a Study in Economics and Politics* (Allen & Unwin: London, 1983).

[12] See Morris, D., 'Cooperative agreements in international business: opportunities and obstacles', Unpublished mimeograph of a report on a conference held at the Institut superieur de l'entreprise et de l'administration des affaires (INSEAD), Fontainebleau, Feb. 1984.

France was instrumental in creating the European Research Co-operation Action for the civil industries, commonly known as Eureka, as a response to the technological challenge from the USA represented by the Reagan Administration's Strategic Defense Initiative (SDI). This was followed, in June 1989, by the European Co-operative Long Term Initiative for Defence (EUCLID), the result of efforts on the part of the French Government to create a complement to Eureka for the European arms industries. The EUCLID projects are novel in that they focus on applications 'upstream' from ready equipment or systems, emphasizing the development of advanced technology in critical areas. (Eureka projects target the pre-competitive stage of development.) EUCLID projects require the consensus of the foreign ministers or the IEPG National Armaments Directors of the participating countries. In the case of senior partnerships, the conferring of leading nation status and the allocation of sub-contracts also require consensus. The planned 1990 allocations for EUCLID total about 120 million European Currency Units (ECUs, roughly $150 million); one-third is provided by the French Government. This budget represents a nominal 2 per cent of the total military R&D expenditure of the IEPG member states.

Two problems associated with EUCLID remain unresolved. First, although the EUCLID agreement stipulates that each participating country should have free access to intellectual property rights to technologies developed within the programme, the procedure for such access has not been defined. Second, while the participation of countries with developing defence industries (DDI) is encouraged, the form of their involvement has yet to be established. While the financial participation of DDI countries is welcome, the leading participants oppose the transfer of all but a few of the new technologies developed within programmes in which DDI nations participate. DDI nations are thus concerned that their role will be limited to a financial one, and that they will not benefit technologically and industrially from their investments.

Another type of co-operation that French firms are seeking with other European firms involves the establishment of binational joint ventures for the production of specified products or systems. For example, three important French arms producers firms announced the creation of such joint ventures in 1989–90:[13]

1. At the end of 1990 Aérospatiale and the German firm MBB merged their helicopter divisions into a new company, Eurocopter S. A., with Aérospatiale holding 60 per cent and MBB 40 per cent of the shares. The new firm is expected to reach a sales volume valued at 10 billion francs ($1.8 billion) in 1990, representing 15 per cent of the world helicopter market, the USSR and the former socialist countries in Eastern Europe excluded.

2. Following several years of negotiations, Aérospatiale and Thomson-CSF in June 1989 entered into a joint venture with the Italian firm Selenia. The new

[13] The largest of the planned international co-operative ventures, involving Thomson-CSF and British Aerospace, did not come about. The planned Eurodynamics project was cancelled in early 1991 because of changed perceptions, on the part of the two co-operating companies, about future market prospects.

Table 8.1. Number of French and British arms co-operation agreements, 1988–89

Agreements are listed by number of partners and by country of ownership of co-operating firms.

By number of partners	France	UK	By country	France	UK
One partner	17	8	FRG	20	9
Two partners	5	5	Italy	6	8
Three partners	3	3	USA	3	6
Four partners	1	1	Netherlands	2	3
More than four partners	1	4	Spain	1	6
Total	**27**	**21**		**27**	**18**

Sources: Adapted from Groupement des Industries Françaises Aéronautiques et Spatiales, *L'industrie aéronautique et spatiale française* (GIFAS: Paris, 1988); 'Exposé des prévisions budgétaire du Ministère de la Défense du Royaume-Uni pour l'anneée 1989' (British Ministry of Defence: London, 1989).

firm, Eurosam, was set up with the specific aim of producing three missile systems for the French and Italian armed forces. The two governments together contribute 9.5 billion francs ($1.5 billion) to the venture.

3. In December 1989 Matra and the British firm GEC merged their space divisions to form a new company, Matra Marconi Space (NV). Although far behind Hughes Aerospace, with 7 billion francs ($1.1 billion) in sales, the new firm ranks in second place among world satellite manufacturers, with 3 billion francs ($500 million) in sales. Also in 1989 Matra entered into a joint venture with MBB. This new firm, Eurodrone, was set up specifically to produce a drone developed through joint research efforts dating back to 1985.

This type of co-operation differs in at least two ways from the type that the French arms industry has pursued in the course of the past 20 years. First, the rise of joint ventures marks an end to the leadership role that French firms may seem to have maintained in past co-operation arrangements, as illustrated in the figures for 1988–89 (table 8.1). These figures invite two observations to be made on the past pattern of French arms co-operation:

1. In order to maintain a dominant role in each co-operation project, French firms have consistently preferred single-partner agreements. By contrast, for the British industry, in the 1980s the most similar to the French in terms of size and presence, a majority of the co-operation agreements were multilateral.

2. French firms have mainly pursued co-operation agreements with German firms, while the UK has maintained a more diversified approach. Constitutional constraints on German arms production have allowed French partners in such ventures to benefit from German financial and technological resources while at the same time maintaining a leadership role in production. As a consequence of the strict rules on German arms exports prohibiting sales to belligerent nations, imposed by the Allies at the end of World War II, French firms have also been

able to reap the main share of the profits from exports of co-produced arms. This leadership role is now being challenged by a restructured and vibrant German industry, which has caught up with, and some say even surpassed, its French counterpart. This applies in particular to the aeronautical sector. The implications for France are discussed in section V below.

Second, this new type of co-operation differs from the the older type in that it is more institutionalized, characterized by asset mergers and the creation of new companies 'upstream' from arms development and production collaboration. The effective use of such formal instruments has resulted in the emergence of an oligopolistic division of the European arms market between a few large consortia, dominated by the leading firms in each country and field. This institutionalized framework for co-operation means that in order for these multinational conglomerates to expand in the international arms markets, they still rely on, and are conditioned by, the influence exerted by the states to whose corporate laws they owe their existence. In effect, financial, technological and (once they export their goods) commercial ties create among the co-contracting firms a union of interests—but notably only in the markets targeted by the joint ventures. Co-operation in one area of production does not rule out competition in other arms (as well as civilian) markets. The interests of each firm participating in international co-operation arrangements are conditioned by the historical, financial and personal links to the country in which it is based; whether advantageous or disadvantageous, these constitute the basis for each firm's effort to penetrate world markets and participate in joint ventures.

V. Consequences of European co-operation for the French arms industry

A peculiarity of French arms industrial groups, whether public or private, is that their might relies heavily on their close relationship with the state. While some members of the French arms oligopoly have branched off into civil production (often dependent on the public procurement market), even such diversified firms rely for their revenues mainly on generous defence contracts and export subsidies, which explains why few industrial groups have ever withdrawn from the lucrative arms market. The opposite seems rather to be the case, namely that large producers on the civil side seek to enter it, as illustrated recently by the cases of two major producers: Alcatel, wishing to re-enter the military telecommunications market which it had left in 1983 (following a 'historic' accord with Thomson described by the French press as the 'Yalta' of electronics); and Framatome, which in order to compensate for its losses in the declining nuclear industry has recently acquired large facilities in the USA and in France to become a major player in the flourishing arms connectors market.[14]

[14] Connectors is a collective term for electrical plugs, sockets and wires. For Framatome's diversification into this market, see Dawkins, W., 'Framatome prepares a strategy for survival', *Financial Times*, 27 July 1990.

In this context, the dominant position in the national economy of groups such as Thomson, Aérospatiale and Matra can only be reinforced as a consequence of the trend towards co-production ventures with other European firms. Declared 'national champions' by the Ministry of Defence, these leading groups benefit from the public financial and technological resources made available in support of European industrial integration. Over the years, Government resources are thus increasingly funnelled to a smaller number of recipients. Moreover, as a result of these co-operation arrangements, the dominant French firms gain increasing influence over the state agencies that are to regulate their activities. The supervision and auditing of projects and cost overruns become more difficult; or to be more precise, while the state auditors know in advance that costs are likely to exceed initial estimates, they accept this as an inevitable part of the national arms procurement process. This means that France's military budget is likely to remain on a high level and that arms production costs will continue to rise.

Will France succeed in defraying these costs through increased sales on the European arms market? Nothing is more uncertain. Given that governments are the sole buyers, it is likely that contracts will continue to be allocated to national firms. Some minor market share gains may be made in the course of the effort to achieve a common European arms market.

The second problem is how the French arms industry will manage the new strategy based on European collaborative agreements. Since the 1960s the industry has mainly based its production and market strategies on co-operation among French companies, and only to a minor extent on co-operation with foreign firms. As discussed above, French firms may not expect to play a leadership role in future European arms co-production ventures. France's presence in the European arms market is weak not only in comparison to that of the USA, which dominates the world arms market to an extent no longer conceivable in civil markets, but also in comparison to that of its main European rivals, Germany and the UK, which have been able to profit from their NATO memberships (see appendix A).

This weakness is in part due to the fact that the French arms industry is spread thin across many sectors, rather than consolidated in a few, internationally competitive groupings. This is a consequence of the past policy of the DGA to plan production according to the defence needs of the state rather than the economic needs of the industry. In practical terms this means that the DGA in the past has not only allocated Government contracts to the leading producers but has also subsidized the efforts of middle-sized firms to develop an expertise in certain niche technologies. The arms industry has thus become one of the few in France for which there is a Government industrial policy to speak of, aimed at maintaining a dense and relatively coherent industrial system integrating large and middle-sized companies.

As argued in this chapter, this policy will have to change. One of many ideas circulating within the Ministry of Defence is that in order to cope with British,

German and even Italian competitors, the French arms industry should be reorganized around two large groups, led by Thomson and Aérospatiale, responsible for defence electronics and aeronautics, respectively. However, this solution would not be acceptable to the shareholders of the other leading French arms firms, such as Dassault and Matra.[15] The latter are hardly prepared to lose their share of the lucrative arms market or the close relationship that they have nurtured with state agencies. The same applies to the state-subsidized middle-sized firms, even more menaced by the restructuring of the arms industry. One cannot outright reject the hypothesis that the hunt for US partners on the part of French firms, and their increasing recourse to co-operation agreements with other European companies, is a means for them to gain sufficient economic significance to maintain their standing in the French arms oligopoly and thus the protection of the Ministry of Defence. However, this naturally self-interested attitude, although now accepted and supported by the Ministry of Defence, has little to do with the arms industrial policy that shifting governments have consistently pursued since the 1960s. Thanks to its relative coherence and to a brief sales boom to the Third World, this policy veiled—and to a degree even mitigated—the predatory effects of the national arms industry on France's productive resources and the potential of its skilled labour force.

[15] Interview with Matra's Executive Director, M. Lagardère, 'Moi vivant, Matra ne se débarassera jamais de ses activités défense', *Echos*, 23 Nov. 1989.

9. Consequences of the French arms exports crisis

Jacques Fontanel and Louis Pilandon

I. Introduction

Although arms transfers constitute a significant component of world trade, they have received relatively scant attention in economic analyses. This is due partly to the difficulty in securing reliable data, and partly to the predominance of strategic and political interests over economic considerations.[1] Classification is another source of difficulty. It is at times unclear whether a transaction involving dual-use technology, such as aircraft or electronics components, should be classified as a transfer of civil or military goods. The nomenclature is continually changing, with restricted products progressively becoming so commonplace that they lose their strategic interest and are as a consequence released for sale on open markets. The time-frame is also a critical factor: for arms exports the time lag between order and delivery dates is generally long (for France on average more than two years). Also, the data sources (customs and defence authorities or institutions such as the US Arms Control and Disarmament Agency, ACDA, or SIPRI) do not publish the same figures. In this chapter, data from the French Ministry of Defence are used.

Furthermore, in the arms market the transaction price is rarely well defined.[2] The transfer takes place as part of a package involving equipment, spares, training, access to technology, export credits, insurance for payment, offset agreements and counter-trade arrangements. Hence, the national export figures are very difficult to analyse. The net costs or revenues to the countries concerned may be different from the nominal prices. Moreover, the figures do not represent effective payments; they stand for a total 'value', and it is common that the importing country is unable to pay the debt. In this case, payment is guaranteed by the government of the exporting firm, that is, the arms transfer becomes an instrument of export promotion.

Of particular interest to the French industry, Third World countries comprise a major source of demand for arms traded on the international market. In the 1960s and 1970s arms transactions became more commercial, as the oil revenues of the Organization of the Petroleum Exporting Countries (OPEC)

[1] Smith, R., Humm, A. and Fontanel, J., 'The economics of exporting arms', *Journal of Peace Research*, vol. 2, no. 3 (1985), pp. 239–47.

[2] Fontanel, J., 'French arms industry', *Cahiers du CEDSI*, no. 10 (Université des Sciences Sociales de Grenoble: Grenoble, 1990).

Table 9.1. French arms export orders and deliveries, 1970–90

Figures are in billion French francs, at current and constant (1986) prices.

	Orders		Deliveries		Deliveries as share of total arms production (%)
Year	Current	Constant	Current	Constant	
1970	..	..	2.7	11.3	*19*
1971	..	..	3.0	11.9	*20*
1972	..	..	4.0	15.0	*24*
1973	..	..	5.2	18.4	*26*
1974	18.3	60.4	6.7	22.1	*30*
1975	16.5	47.9	8.3	24.1	*32*
1976	18.9	49.1	11.6	30.2	*37*
1977	27.4	64.9	14.7	34.8	*41*
1978	21.7	47.0	17.3	37.5	*40*
1979	25.1	50.0	20.5	40.8	*40*
1980	37.4	67.0	23.4	41.9	*40*
1981	33.8	53.4	28.5	45.0	*41*
1982	41.6	57.8	28.9	40.2	*38*
1983	29.1	36.4	33.1	41.3	*38*
1984	61.8	70.5	41.9	47.8	*42*
1985	44.5	47.1	43.9	46.5	*42*
1986	25.3	25.3	43.1	43.1	*40*
1987	28.9	27.9	34.1	32.2	*30*
1988	37.5	35.4	38.2	36.1	*32*
1989	20.0	18.0	37.3	33.6	*31*
1990	37.0	33.0	..	..	..

Sources: French Ministry of Defence; Lewis, J. A. C. and Reed, C., 'France fights the export battle', *Jane's Defence Weekly*, 25 May 1991, p. 893.

provided an alternative source to finance purchases. Alongside these quantitative changes, this period also saw important qualitative changes in demand. While the arms transferred to the Third World in the 1960s were largely obsolete, outdated or second-hand, starting in the 1970s the major exporters in the industrialized world began to sell modern weapon systems to Third World buyers. This change is partly a consequence of the aggressive commercial policy of French private and public military enterprises.

In the course of these developments, the international sale of arms and technology was progressively detached from foreign policy and strategic objectives. As economic interests began to take precedence, importing countries encountered fewer political hindrances to buying high-technology conventional arms. For the French Government, which sought to maintain a national arms industry that could continue to ensure independent access to the latest military technologies, export sales at prices above short-run marginal costs made some contribution to investment costs. In political terms this economically oriented arms export policy still provided France with an instrument to influence the policies of purchasing governments and to assist friends and allies.

Table 9.2. Value of French arms export orders and deliveries, by region, 1985, 1987, 1989

Figures are in billion francs, at current prices.

	1985		1987		1989	
Regions	Orders	Deliveries	Orders	Deliveries	Orders	Deliveries
Europe and North America	18.5	6.2	11.8	8.5	9.0	11.8
North Africa and the Middle East	17.5	26.4	11.3	18.4	5.5	20.9
South America and the Caribbean	3.3	2.1	2.0	2.0	1.5	1.7
East Asia	2.4	7.0	2.9	4.0	2.8	2.0
Sub-Saharan Africa	2.0	1.6	0.6	0.9	0.9	0.6
Other	0.8	0.6	0.3	0.3	0.2	0.3
Total	**44.5**	**43.9**	**28.9**	**34.1**	**20.0**	**37.3**

Sources: *Avis présenté au nom de la commission de la défense nationale et des forces armées sur le projet de loi de finances pour* [year], *Tome V, Défense, Recherche et industrie d'armement*, reports for 1988 (no. 963), 1989 (no. 923) and 1991 (no. 1638) (Assemblée Nationale: Paris, 8 Oct. 1987, 12 Oct. 1989 and 10 Oct. 1990, respectively).

The 1980s were the beginning of a global trend towards the production of 'appropriate technology' arms, cheaper and better tailored to suit Third World needs. The exporters were increasingly found among newly industrialized countries. Partly as a result of this, by the end of the 1980s a crisis had developed in the international arms market, and the French arms industry was directly affected by a reduction in orders. The main solution pursued by the Government is increased international co-operation, mainly with European Community (EC) countries (see chapter 8).

II. French arms exports: facts and figures

Although 1987 was a year of increasing international arms sales, French arms exports fell by 18.6 per cent as compared to figures for 1986, with exports to developing countries falling by 14 per cent and to industrialized countries by 50 per cent. French arms export orders were valued at 61.8 billion francs in 1984, 44.5 billion francs in 1985 and 25.3 billion francs in 1986 (see table 9.1). This drastic decline was mainly due to the gamble on the part of French producers to invest in the US arms market just as the US Government began to reduce the growth of its military expenditure, as a result of congressional action to seek to stem a growing public deficit. Other factors included the continued impoverishment of French Third World customers, the fall of the dollar exchange-rate and growing international competition. Arms exports of West European countries have shown a fluctuating downward trend since 1987, but French exports have

Table 9.3. Value of French arms export orders and deliveries, by equipment category, 1985, 1987, 1989

Figures are in billion francs, at current prices.

	1985		1987		1989	
Equipment category	Orders	Deliveries	Orders	Deliveries	Orders	Deliveries
Air	28.2	22.1	15.5	18.2	10.4	20.4
Land	11.6	13.6	7.9	12.7	7.1	13.9
Naval	4.7	8.2	5.5	3.2	2.5	3.0
Total	**44.5**	**43.9**	**28.9**	**34.1**	**20.0**	**37.3**

Source: *Avis présenté au nom de la commission de la défense nationale et des forces armées sur le projet de loi de finances pour* [year], *Tome V, Défense, Recherche et industrie d'armement*, reports for 1988 (no. 963), 1989 (no. 923) and 1991 (no. 1638) (Assemblée Nationale: Paris, 8 Oct. 1987, 12 Oct. 1989 and 10 Oct. 1990, respectively).

been particularly affected.[3]. It is interesting to note that while French arms sales abroad have declined sharply, exports are not yet critically low, as deliveries lag behind orders (see tables 9.2 and 9.3). However, given that orders usually are greater than deliveries, the French arms industry may be expected to suffer serious setbacks, as was the case in 1990 with exports to Iraq.

An arms exporting country with an effective monopoly is able to exact a high price, in political as well as economic terms, for a desired weapon system. However, this is rarely the case for France, which in each category of arms encounters competition from other countries. During the 1970s French firms made skilful use of the country's relative independence from the two superpowers, and of the weak government control over arms exports, to carve out a significant share of the international arms market. During the 1980s this advantage was reduced with the emergence of new suppliers. As a result, the competitive position of the French arms industry is declining. According to an independent calculation, in 1984 France exported 50 per cent of the arms it produced, as compared to 20 per cent for the FRG, 70 per cent for Italy and 42 per cent for the UK.[4] Since 1984, this share has declined (see table 9.1).

Arms production is characterized not only by initially high overhead costs, owing to outlays for research and development (R&D), but also by declining costs in later stages, as a result of rising learning curves (increased efficiency through experience) and economies of scale. As large firms can produce more cheaply, arms exports become for almost all countries an economic condition for an efficient national arms industry. French companies would not bid for contracts if they did not expect them to be profitable, and profitability often

[3] Anthony, I., Courades Allebeck, A., Hagmeyer-Gaverus, G., Miggiano, P. and Wulf, H., 'The trade in major conventional weapons', SIPRI, *SIPRI Yearbook 1991: World Armaments and Disarmament* (Oxford University Press: Oxford, 1991), table 7.1, p. 198.

[4] Brzoska, M. and Ohlson, T., 'The trade in major conventional weapons', SIPRI, *World Armaments and Disarmament: SIPRI Yearbook 1986* (Oxford University Press: Oxford, 1986), p. 336.

Table 9.4. French arms production and export, 1980, 1984, 1987

Production figures are in billion francs, at current prices; export figures are percentage shares of production.

	1980		1984		1987	
Sector	Production	Export	Production	Export	Production	Export
Electronics	14.9	*44*	25.6	*47*	34.1	*33*
Aeronautics and space	22.8	*45*	34.7	*47*	33.1	*39*
Shipyards	4.9	*27*	10.5	*43*	9.0	*21*
Land armaments	10.1	*47*	17.1	*49*	18.2	*38*
Nuclear	3.1	*0*	5.5	*0*	5.8	*0*
Other	2.7	*16*	4.9	*16*	6.8	*15*
Total	**58.5**	..	**98.3**	..	**107.0**	..

Source: French Ministry of Defence.

depends on subsidies (especially for R&D) from the contracting government, credits and official aid, and approval for exports. The benefits of arms exports appear directly and accrue to particular interests. The costs are less obvious and depend on the alternative use of funds.

The advantages of exports are often difficult to assess. The sale in 1985 by the Mirage 2000 consortium—consisting of Dassault, SNECMA, Thomson-CSF and Matra—of 40 fighter aircraft to the Greek Government was conditional on a more general agreement on industrial investments, including purchases of products and services. For a contract of 8 billion French francs ($890 million), the four firms must offer 4.8 billion francs ($536 million) in compensation. Of this, 1 billion francs ($111 million) was to have been paid from June 1985 to June 1989. In other words, it is not clear that the French economy in the end will benefit from this programme.

While there thus are powerful economic incentives to export arms, such exports do not necessarily contribute to the national economy. Arms production requires that scarce scientific and technical resources be diverted from potential civil uses that could improve the productivity and competitiveness of the economy as a whole. To this must then be added the additional resources invested in effecting exports. In France, this growing dependence of the arms industry on exports has created a powerful economic lobby—a coalition of industry, trade union, and national and regional political interests—for continued large-scale government subsidies and compensations, despite the fact that such measures lack any logic in terms of national economic priorities.

Econometric models using recent data suggest that military expenditure has both positive and negative effects on arms exports.[5] When arms export forecasts suggested the emergence of a crisis for industry, military expenditure was increased in order to compensate industry for the lack of demand. The French

[5] Smith, Humm and Fontanel (note 1).

Government's view on arms exports was usually determined by national defence equipment needs, with the aim in sight to reduce procurement costs by maximizing production series, and thus sales. However, the effort on the part of the Government during the 1980s to modernize French armaments, through an exceptional increase in equipment spending relative to operating costs, was rather due to pressure from the French arms lobby, which used unemployment forecasts and figures on export decline and the need to maintain the competitiveness of the industry to argue its case. By increasing the procurement budget at least until the 1990s, the Government has compensated the French arms industry for its losses on the international markets.

In order to explain the evolution of French arms transfers in the late 1980s, it is useful to examine whether empirical links exist between the percentage of arms transfers in arms production, the annual rate of growth of 'Titre 5' of the national military budget (capital expenditure) and the percentage of equipment spending in the military budget. These relations are very complex, but it seems that the coefficients tend to be negative in the short run because of crowding-out effects, and positive in the long run because of the economies of scale in national arms production. While the empirical results are not clear cut, what very tentative conclusions can be drawn from this? Perhaps a link can be shown to exist between equipment spending and arms transfers during a given year, but the regression results for equipment spending and 'Titre 5' are not good.

Arms exports are in France usually regarded as a complement to national defence requirements. As noted above, in the 1980s the Government compensated for the losses to the arms industry of international market shares by increasing national procurement. However, the effect of this policy has not been confirmed by econometric results, in part because the 1977–89 period is heterogeneous. During this period the international arms market, originally dominated by suppliers, gradually became dominated by buyers.

A 1988 parliamentary report on the French military industries explained the export crisis as a result of buyers having less to spend, the emergence of new producers and the weakness of the French system of arms sales, characterized by financial rigidity and ageing marketing companies. The French system for export sales is contrasted with its British counterpart, with the British barter agreements with Saudi Arabia (involving the sale of a number of Tornado aircraft) and Malaysia (a package deal involving, among other things, the sale of a number of Hawk fighter/trainer aircraft) given by way of examples. The report recommends improvements in the financial and insurance control structure.[6] For instance, as part of a sales agreeement, the Government will

[6] 'I note that in contrast to their British colleagues, French political leaders have always felt strong reticence towards the idea of extracting a certain amount of commercial benefit from certain of our diplomatic positions. In general, ministers of defence seem to be more sensitive to this issue, but in other countries they are unfortunately perceived as being technical ministers above all. It would no doubt be desirable that those ministers that appear more political in the eyes of potential buyers also share their preoccupations.' Branger, J.-G., *Avis présenté au nom de la commission de la défense nationale et des forces armées sur le projet de loi de finances pour 1989, Tome V, Défense, Recherche et industrie d'armement* (Assemblée Nationale: Paris, 13 Oct. 1988), p. 17 (authors' translation).

provide French technical personnel to assist in the design of an Indian aircraft-carrier, to be produced in Cochin.

In the arms market as in any other market, an effective demand requires that means of payment are available. Without financial guarantees, it is in the long term difficult to maintain arms transfers in periods of economic decline, unless military assitance programmes cover the cost, as is often the case in the USA. The parliamentary report nevertheless pleads for a new export-oriented product policy, promoting quality research, a wider geographical market, greater company control over foreign acquisition, improved risk insurance for the military sector and further European industrial collaboration. It should be noted that this is a political, not an economic, policy.

As the costs of production and procurement grow, the significance of arms industry co-operation continues to increase. Under such economic conditions, it becomes difficult to maintain a course dictated by the strategic concern to maintain an independent supply of arms.

III. Armament co-operation

The typical pattern for armament co-operation is that development costs are shared among partners. This amounts to savings to each only if the defence organizations of the countries involved need exactly the same systems. French arguments against armament co-operation revolve around the ideas that: (*a*) French weapons are superior and are tailored exactly to the needs of the French armed forces; (*b*) a domestic arms industrial base is essential to maintain strategic independence; and (*c*) 'unfair trade' practices motivate protective measures. On the economic side, it is further argued that (*d*) domestic procurement creates employment, boosts tax revenue, improves the balance of payments and produces technological spin-offs for civil production; (*e*) if co-operating countries do not want exactly the same systems, in meeting the needs of each partner additional costs are incurred which may offset the advantages of large-scale production; (*f*) compromise designs may be more expensive to produce; and (*g*) co-ordination and transport needs increase costs, especially as collaboration schemes commonly involve complicated, and politically rather than economically motivated, work-sharing and compensation arrangements.

The main arguments for armament co-operation stress: (*a*) the difficulty to develop within a single enterprise the specialized, high and varied technical competence needed to maintain a dynamic arms sector; (*b*) the need to reduce R&D investment costs per firm; (*c*) the need to spread the substantial risks involved; (*d*) the possibility to reduce unit costs by increasing production series; and (*e*) the possibility of increasing sales by pooling export markets.

A policy of national self-sufficiency and independence in arms production can prove both expensive and dangerous. For cost reasons, France will be forced to follow the path towards European arms co-operation and specialization. The alternate course of maintaining production capacities by increasing procurement and thus defence spending would place a heavy burden on the national economy and lead to a decline in industrial productivity which, in the long run, would reduce growth opportunities, thus threatening national security.

The Délégation Générale pour l'Armement (DGA) directs France's military policy with the declared twin goals of independence and solidarity. Independence refers to autonomy in decision making. The aim is to harness national energies and skills to provide the foundation for a defence based on France's own resources. By solidarity is implied that once autonomy is assured, France should collaborate with its allies, at least in the design and introduction of weapons needed for their common security. The decision to develop the arms industry seems primarily to be an effort to satisfy the requirements for national independence.

Economic considerations nevertheless define the limits of this arms industrial policy. Besides being guided by economic principles to make the most efficient use of investments, this policy also involves efforts to involve the arms sector in the modernization and industrialization of the French economy. In this greater context, however, the requirements of a national arms industrial policy raise special problems, mainly in terms of the need of the arms industry for constant technological advancement and for high—and thereby risky—investments.

Co-production is one way to address these challenges. In the mid-1970s the EC countries began to lose ground in high technology. From 1975 to 1985 imports increased by 8 per cent while exports declined by 2.5 per cent. In the face of a growing technological gap there was an apparent need for European co-operation in research, not the least in areas of military technology. Although programmes such as the European strategic programme for research and development in information technology (Esprit), basic research in industrial technologies for Europe (Brite) and Eureka were established to promote co-operation in civil research, these have also had, and continue to have, relevance for military R&D. In addition to these, the Independent European Programme Group (IEPG) favours strenghtening co-operation for the structural rationalization of European arms production resources. Such institutional frameworks are discussed in chapter 3.

In terms of specific programmes, bilateral arrangements seem to have had greater success than multilateral initiatives, for France as well as for other EC countries. French firms have signed agreements with a number of foreign companies. Table 9.5 shows French involvement in the major West European armament co-operation programmes of the 1980s. The success of such co-production projects has depended on common requirements, a well-structured industrial and government-to-government organization, a dependable cost–benefit analysis of the advantages/disadvantages of co-operation, an improvement in or maintenance of the competitiveness of the national industries involved in the programme and good prospects for exports.

Table 9.5. Major French armament co-operation programmes, 1977–87

For programme and firm acronyms, see pages xiv–xv; net sales are in billion francs, at curent prices.

Weapon type/ programme	Financing Country	%	Firms	Orders Country	No.	Year of delivery	Techno-logy sales	Net sales
Missile	France	*50*	Aérospatiale	France	17 000	1974–91	None	2.1
HOT	FRG	*50*	MBB	FRG	25 000	and		
				Other	27 000	beyond		
Missile	France	*50*	Aérospatiale	France	64 000	1974–91	UK	3.8
Milan	FRG	*50*	MBB	FRG	90 000		Italy	
				Other	62 000		India	
Missile	France	*50*	Aérospatiale	France	7 000	1977–88	USA	10
Roland	FRG	*50*	MBB	FRG	14 000			
				Other	3 500			
Mine-sweeper	France	*33*	DCN	France	10	1983–89	None	2.5
Tripartite	Belgium	*33*	Mercantile & Béliard	Belgium	10			
	Netherlands	*33*	Van der Giessen	Netherlands	15			
Helicopter	France	*31*	Aérospatiale	France	40	1978–84	None	2
Lynx			Turboméca					
	UK	*69*	Westland	UK	218			
			Rolls Royce	Other	71			
Helicopter	France	*74*	Aérospatiale	France	343	1973–90	None	2.2
Gazelle	UK	*26*	Turboméca	UK	282			
			Westland	Other	426			
			Rolls Royce					
Helicopter	France	*92*	Aérospatiale	Other	247	1969–89	Romania	2.7
Puma			Turboméca				Indonesia	
	UK	*8*	Westland					
			Rolls Royce					
Communica-tions	France	*95*	Thomson	France	1	1981–85	None	8.3
RITA	Belgium	*5*	Bell	Belgium	1			
				USA	1			
Aircraft	France	*50*	Dassault	France	200	1972–82	None	9.2
Jaguar			Turboméca	UK	203			
	UK	*50*	BAe	Other	94			
			Rolls Royce					
Aircraft	France	*50*	Dassault	France	175	1978–84	None	7.8
Alpha Jet	FRG	*50*	Dornier	FRG	175			
				Other	15			

Sources: Bechter, J.-P., *Avis présenté au nom de la commission de la défense nationale et des forces armées sur le projet de loi de finances pour 1988, Tome V, Défense, Recherche et industrie d'armement* (Assemblée Nationale: Paris, 8 Oct. 1987), p. 35; de Villepin, X., *Avis présenté au nom de la commission des Affaires étrangères, de la défense et des forces armées sur le projet de loi finances pour 1988, Tome IV, Défense Section Commune* (Senat: Paris, 1987).

There are also management companies that on a competitive basis contract out production of products jointly developed by firms in various NATO countries. Examples of consortia are the French–German Euromissile, registered in Paris in 1972, and the French–German–British Euromissile Dynamics, registered in Paris in 1976 (for examples, see table 2.4 in this book). To achieve industrial co-operation with other European arms producers, French firms and the French Government must overcome a number of inherent difficulties, some of a general nature, others specific to French procedure and structures:

1. The basic military needs of the contracting parties may not be identical, whether in terms of strategic interests or time-frames.
2. The French administrative and financial procedures are often not in keeping with those of other countries.
3. Colliding interests lead to disagreements over export policies.
4. Governments tend to support national industries, even when these are not competetive on international markets.
5. The conception and execution of programmes tend to be delayed.
6. Collaboration is costly.

Another challenge is posed by the French trade unions, which claim that European military arms co-operation is the main cause of the decline in manufacturing in France, to the detriment of regions and workers. An example given is the European Space Agency (ESA), which while primarily funded by France, is dependent on co-finance agreements with other participating countries. In return for economic participation, these countries require a transfer of technology that leaves French industry at a relative disadvantage. Similarly, the trade unions regard arms industry co-operation among the EC countries as disadvantageous both in terms of employment and in terms of reaping the benefits of technological advancements and investments.

> West Europeans have relied heavily on collaboration to sustain their military–industrial–scientific–technological system . . . Foreign support is sought not only for specific weapons projects but also for national organizational structures themselves and for their continuing technological advancement. These organizational structures, while indigenously based, depend critically on external technological know-how and on guaranteed markets among collaborating states. The cost of weapons research and development and the high unit costs of exclusively national production, in the absence of available global marketing outlets, generate strong incentives to cooperate. This is so despite inevitably higher average costs of multinational versus national programs. Thus, the markets drives a major portion of Euro-defense production, although individual states seek general political autonomy in arms production as well.[7]

The French arms industry has relied heavily on exports. It must now focus more on international co-operation, since exports are declining.

[7] Kolodziej, E. A, 'The political economy of making and marketing arms: a test for the systemic imperatives of order and welfare', Paper presented at the Meeting of the International Studies Association, London, 1 Apr. 1989.

10. Daimler Benz: the final stage of concentration in German arms production?

Michael Brzoska and Peter Lock

I. Introduction

For most of post-World War II German political history, domestic arms production was a non-issue. Without much political discussion, a sizeable arms industry was built up during the 1950s and 1960s. On occasion the fact that most of the 'new' arms producing firms in fact had traditions in arms production going back to the period before 1945 created a public stir,[1] but in general the public mood was to let decision makers in this sector act without public interference.

Those building up the arms industry also tried to project a low arms production profile. Companies such as Krupp, whose owner Alfried Krupp von Bohlen und Halbach after the war vowed never again to produce weapons,[2] used the term *Sondertechnik* (special technique) for their arms production. The procurement authorities and most politicians stressed the civilian benefits of producing arms and encouraged companies to limit the share of arms production in their total turnover to a few per cent.

Under these circumstances, the political willingness of one particular state of the Federal Republic, Bavaria, reinforced by the political skill of one person, Franz Josef Strauß, could shape the structure and location of West German arms production in a profound way. Strauß, in various periods Minister for Nuclear Matters, Defence and Finance between 1958 and 1969, and his collaborators turned Bavaria—until the 1950s an agrarian, industrially backward state—into an economic area marked by advanced technology and industrial production. The concept of using arms production as a nucleus for a multi-sectoral high-technology industry worked quite well between the 1950s and the 1990s. The extent of arms production in Bavaria, especially in the Munich area, is still greater than in any other economic area of the Federal Republic of Germany of comparable size (see figure 10.1).[3] By far the largest and

[1] For examples of such public debate, see Perdelwitz, W. and Fischer, H., *Waffenschmiede Deutschland* (Gruner und Jahr: Hamburg, 1984).

[2] See Engelmann, B., *Krupp: Die Geschichte eines Hauses* (Goldman: Munich, 1978), pp. 456, 480.

[3] See Huck, B. J., 'The regional and occupational dependence on defense contracting in the Greater Munich area', eds L. Paukert and P. J. Richards, *Defence Expenditure, Industrial Conversion and Local Employment* (International Labour Office: Geneva, 1991), pp. 51–70.

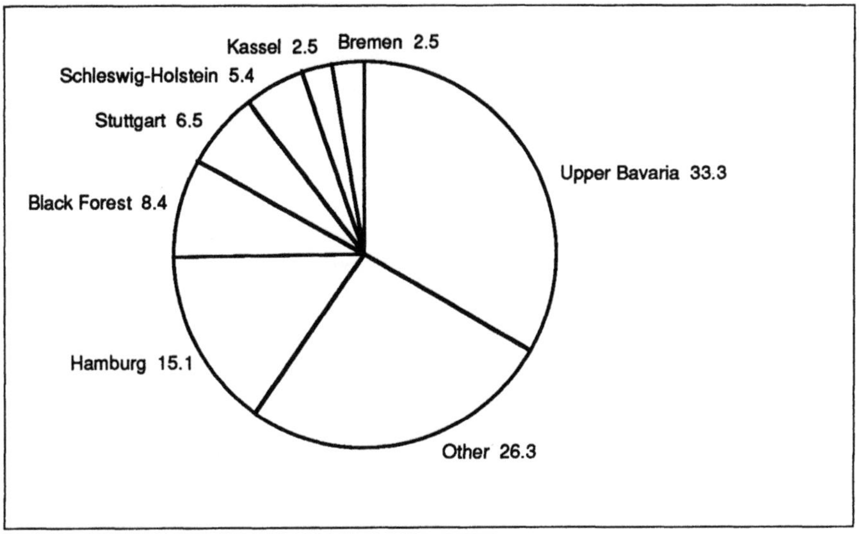

Figure 10.1. Regional distribution of FRG procurement orders, 1989

Figure shows percentage shares.
Source: Deutscher Bundestag, document no. 11/7441, 20 June 1990.

technologically most advanced of these arms producing firms is MBB (Messerschmitt–Bölkow–Blohm).

Between 1987 and 1989, arms production became one of the most intensely debated issues in (then West) German politics. The take-over of the predominantly publicly owned arms producing company MBB by the automobile and truck producer Daimler Benz raised great fears about an 'arms giant', 'a new German arms power' that would dominate Germany today and 'the world tomorrow'.[4] The handling of and the circumstances surrounding the take-over, summarized below, reinforced these fears.

In organizing the sale of MBB, Daimler Benz and the federal Government were only partly successful in their attempts to convince the public that arms production was of little importance in this merger, that Daimler Benz was interested in supplementing its automobile production by diversifying into high technology of all kinds and that the Government was content to get rid of a loss-making state-owned enterprise.

The unclarified circumstances surrounding the merger raise a number of questions. Does it entail a dangerous conglomeration of power, with serious political and military implications? Or is it merely, as the Daimler Benz management has stated, a necessary step for that firm to attain the level of high-technology competence it needs in order to compete on a world scale? What is

[4] See Büschemann, K.-H. and Schmid, K.-P., 'Unter schlechtem Stern: Bonn begibt sich in die Hand des Rüstungsriesen Daimler', *Die Zeit*, 11 Nov. 1988, pp. 25–27; 'Die neue deutsche Rüstungsmacht', *Der Spiegel*, vol. 42, no. 31 (1 Aug. 1988), pp. 24–43.

the outcome? What are the consequences for other German arms producers? Is it a harbinger of monopolization of arms production in Germany or a step towards the 'civilianization' of arms production?[5]

II. The road to the take-over

Negotiations on a possible take-over were initiated by the federal Ministry of Economics in Bonn. Charged with the responsibility to oversee the West German aerospace industry, the Ministry had made MBB the by far largest firm in the sector. One small producer after another was pressured to join MBB. By the mid-1980s only the family-owned Dornier company remained as a sizeable, independent aircraft producer. However, MBB continued to make losses. Production was scattered, regionally and in terms of products, the latter ranging from explosives to space vehicles. It was considered inefficient, and was torn by the diverging interests of its public owners, which included a number of federal states. In addition, there were substantial losses from the production of the Airbus transport aircraft, which MBB had transfered to the Government. In the early 1980s the Economics Ministry concluded that something drastic had to be done. It came up with a simple solution, one that had not been possible before: to transfer MBB, including the Airbus losses, to the private sector.

In an attempt to keep MBB in Bavaria, Franz Josef Strauß, at that time Prime Minister of Bavaria, got the federal Government's consent first to contact Bavarian-based companies, such as the electronics giant Siemens and the automobile producer BMW (see table 10.1). However, neither of these firms showed any interest in acquiring MBB. Federal Minister of Economics Martin Bangemann from Baden-Württemberg finally found a more interested party in Daimler Benz, also from Baden-Württemberg. Especially Daimler Benz Vice-Chairman Edzard Reuter was keen to diversify from autos and trucks into aerospace and electronics, which was precisely what the federal Government now offered. After some infighting the Daimler Benz Board of Directors decided to take the course proposed by Reuter, who became Chairman in the process. He was able to prevail over the conservative opposition on the Board that feared that Daimler Benz's core business would suffer from what it perceived to be a risky move into arms and aerospace. Reuter succeeded because he had the support of Alfred Herrhausen, the Chairman of the Supervisory Board of Daimler Benz and Director of the Board of Directors of the Deutsche Bank, which controlled about 40 per cent of Daimler Benz's capital.[6] As initial steps, Daimler Benz took over Dornier and parts of AEG-Telefunken. The company also increased its shareholding in MTU. Out of these

[5] Forster, M., 'Building a defense giant? Daimler-MBB merger "civilianises" German defense industry', *International Defense Review*, vol. 21, no. 9 (Sep. 1988), p. 1205.

[6] Deutsche Bank's direct ownership at the time amounted to only 28.24 per cent, but it gained additional control through transfers of voting rights by small shareholders. The second largest individual shareholder was Kuwait, with 14 per cent.

Table 10.1. Chronology of the creation of Deutsche Aerospace AG (DASA)

Summer 1984	Daimler Benz Vice-Chairman Edzard Reuter presents a concept for the diversification of the company into aerospace to the Board.
Early 1985	The Board of Directors of Siemens decides not to pursue negotiations with the aim to acquire MBB.
Spring 1985	Daimler Benz acquires 100 per cent of the capital of MTU. Daimler Benz acquires majority of the Dornier group.
Summer 1985	After a thorough analysis, the management of BMW rejects an offer to negotiate the acquisition of MBB.
Oct. 1986	Daimler Benz acquires majority of shares of AEG.
Spring 1987	Federal Economics Minister Martin Bangemann invites the Daimler Benz management to take over MBB.
Nov. 1988	The federal Government agrees to the take-over and to the additional financial subsidies to be paid to Daimler Benz.
Dec. 1988	The Board of Directors of MBB agrees to the take-over.
Apr. 1989	The Federal Cartel Office forbids the take-over.
May 1989	Daimler Benz unites MTU, Dornier and parts of AEG (under the new name of Telefunken Systemtechnik or TST) under the new roof of DASA. Daimler Benz applies for ministerial approval of the take-over.
Aug. 1989	The independent Monopoly Commission recommends that the Minister of Economics approve the take-over.
Sep. 1989	The Minister of Economics approves the take-over under certain conditions.
Nov. 1989	The owners of MBB (both state governments and private companies) agree to the take-over, including the new conditions.
Feb. 1990	DASA, with MBB, is reorganized into five divisions: Aerospace (parts of MBB and Dornier), Space (parts of MBB and Dornier), Defence and Systems (parts of MBB, Dornier and TST), Propulsion (MTU) and Other Activities, later named Diversification (parts of MBB, Dornier, TST and MTU).

Sources: Daimler Benz and MBB press and company reports; authors' files.

three companies it created Deutsche Aerospace (DASA)—a misnomer as long as it did not include the largest German aerospace company, MBB. The take-over of MBB was the logical next step of a Daimler Benz company that in a very short time, by 1987, had become an important force in aerospace and arms production.

The possibility of a take-over of MBB by Daimler Benz became a political issue in 1987, as concern about implications was voiced by a wide spectrum of political actors in the FRG as well as abroad. Some critics recalled Germany's militaristic past, alluding to the fatal political coalition between 'monopoly capital' and German militarism and expansionism that had led to two world wars.[7] Others insisted that the power relationship between the state, procuring weapons for defence, and the future predominant sole-source supplier would be reversed, raising the spectre of a dominant military giant that the Government would no longer be able to control.[8] Last but not least, there were those who argued that the sheer size of Daimler Benz (which with the inclusion of MBB would be the largest industrial company in the FRG), as well as its dominance in various market segments, would not be compatible with maintaining a sound market economy.[9]

These grand criticisms overshadowed a number of other issues that played an important role in the actual negotiations. The federal Government did not itself own any shares in MBB, the main owners being the state governments of Bavaria, Bremen and Hamburg. These had local interests to maintain. The labour force had also acquired a strong position in the company. A demand to guarantee the continued operation of all plants was one of the sensitive issues in the negotiations between state governments and Daimler Benz, which was not willing to take all financial risks at once. As a result of these negotiations, the company obtained guaranties for Airbus subsidies and promises of Government military orders extending into the beginning of the next century. These amount, according to one source, to a capitalized value of more than 60 billion DM.[10]

To pursue the take-over, Daimler Benz needed the clearance of the Federal Cartel Office (Bundeskartellamt). In their briefs to the Office, both Daimler Benz and the federal Government (represented by the Defence Ministry) stressed that Daimler Benz must not be measured in national but in West European terms. Daimler Benz argued that it was competing in a European-wide, if not world-wide, market. The Defence Ministry also pointed at the need to look beyond national borders. However, led by doubts that the arms market is as supranational as the architects of the take-over claimed it to be, and relying on a narrow interpretation of its legal brief as supervisor of German market structures, the Federal Cartel Office issued a clear negative verdict:

In the area of arms production it can be expected that the merger would increase the dominant position of the Daimler Benz/MBB group in the markets for military aircraft, helicopters, missiles and propulsion units, and would result in a dominant position in the markets for drones and military electronics. The merger would result in the creation in the armaments sector of a dominant company group whose market influence could not be controlled by favouring foreign suppliers, because of the specific conditions of the armaments economy. This market influence, which in many areas would be the result of a monopoly position as systems leader, would also allow

[7] See Hamburger Stiftung für Sozialgeschichte des 20. Jahrhunderts, *Das Daimler-Benz-Buch: Ein Rüstungskonzern im 'Tausendjährigen Reich'* (Greno: Nördlingen, 1987); Sörgel, A., *Daimler-Benz: Der Multi im Musterländle* (Progress: Bremen, 1986); Bamberg, S. *et al.*, '*. . . und morgen die ganze Welt': Daimler-Benz—ein Rüstungskonzern auf dem Weg ins 21. Jahrhundert* (KOMZI: Frankfurt, 1990).

[8] See Lock. P. and Wilke, P., 'IG Rüstung unter einem guten Stern', *Sicherheit und Frieden*, no. 4 (1987), pp. 243–47.

[9] Roth, W., 'Der größte ordnungspolitische Sündenfall', *Wirtschaftsdienst*, no. 12 (1988), pp. 602–3.

[10] 'Die neue deutsches Rustungsmacht' (note 4).

Table 10.2. Federal Cartel Office estimates of the effects of the Daimler Benz–MBB take-over on the German arms market structure

Market	Share in federal procurement orders (in per cent)	Market position before take-over	Market position after take-over
Fixed-wing aircraft	MBB: 85	MBB overwhelming; Dornier strong	Monopoly on systems level
Helicopters	MBB: about 95	Monopoly	Monopoly
Propulsion	MTU: 83	MTU dominant	Increased dominance of MTU because of direct link to dominant system producers
Missiles	MBB: 80	MBB dominant	Dominance increased because Dornier disappears as competitor
Drones	Dornier: 86 MBB: 14	Dornier dominant	Monopoly
Military electronics	AEG/Dornier: 42 MBB: 17 Siemens: 12.5 Others: 28.5	AEG/Dornier strong but not dominant; MBB important competitor	Daimler Benz dominant, partly due to AEG/Dornier/ MBB merger, plus direct link to dominant major system producer
Naval systems[a]	MBB: 36 AEG/Dornier: 26.3	MBB and AEG/Dornier strong competitors	Dominant position

[a] As a condition for the merger, Daimler Benz had to sell its naval interests.

Sources: *Beschluß in dem Verwaltungsverfahren Daimler-Benz AG u.a.*, document no. B 7-35 00 00-U-137/88 (Federal Cartel Office/Bundeskartellamt, 7. Beschlußabteilung: Berlin, 19 Apr. 1989); Jäger, U., Schwegler-Rohmeis, W. and Berger, W., *Rüstung ohne Grenzen* (Verein für Friedenspädagogik Tübingen: Tübingen, 1989), pp. 68–69.

Daimler Benz to restrict competition on the level of suppliers. The Ministry of Defence, as buyer of arms, would be greatly dependent on this company group. The merger also has importance for other markets because of the absolute size of the joint capacities, which would make the consequences of an increased market influence even more severe.[11]

The Cartel Office found that Daimler Benz, with control over parts of AEG-Telefunken, MTU, Dornier and MBB, would have dominating market shares not only in almost all relevant arms markets (see table 10.2) but also in large segments of the space industry and in some markets for trucks.

German merger law permits firms to appeal decisions by the Federal Cartel Office in two ways. One is juridical, the other political. Not surprisingly,

[11] *Beschluß in dem Verwaltungsverfahren Daimler-Benz AG u.a.*, document no. B 7-35 00 00-U-137/88 (Federal Cartel Office/Bundeskartellamt, 7. Beschlußabteilung: Berlin, 19 Apr. 1989), p. 22 (authors' translation).

Daimler Benz chose the political course, by requesting the Federal Economics Minister to overrule the Office's decision. According to Article 24, Paragraph 3 of the Law against Restrictions on Competition ('Gesetz gegen Wettbewerbsbeschränkungen') the Economics Minister can give special clearance (*Ministererlaubnis*) if he or she considers national economic interests to outweigh the costs of reduced competition. Prior to making such a decision, however, the Minister must first consult the Monopoly Commission, an independent, Government-appointed body consisting of five experts on market structures. While the Chairman of the Commission, Ulrich Immenga, took a strong position against the merger,[12] a majority of Commission members recommended that the Minister allow the merger with some reservations. In a rare discussion on the role of banks in arms production, the Commission listed other arms producing companies in which Deutsche Bank had controlling influence, including Krauss-Maffei, Rheinmetall, Siemens and Thyssen. It recommended that a new law be drafted limiting the influence of banks in this sector.[13]

In September 1989 the new Minister of Economics, Helmut Haussmann (Martin Bangemann resigned earlier to become a member of the Commission of the European Community), gave clearance for the take-over, on the provision that Daimler Benz met certain conditions. These included an agreement that Daimler Benz would sell a number of MBB subsidiaries involved in naval production[14] as well as its 12.5 per cent share of the tank producer Krauss-Maffei, and that Daimler Benz and MBB managers would leave posts as members of supervisory boards of other arms producing companies. Daimler Benz accepted the conditions and the deal was formally closed in late 1989.

III. The industrial background of the take-over

When MTU, AEG, Dornier and MBB successively came under control of Daimler Benz, 5 of the 10 largest German arms producers came under a single corporate command. All of them held at least oligopolistic positions in their respective segments of the arms market. While jointly combining a large share in German arms production, the conglomerate is by no means a homogeneous producer of military commodities.

Telefunken (later AEG) military production and research were at the technological forefront during World War II. AEG-Hochfrequenztechnik, in Ulm, resumed this tradition when military production was slowly re-introduced in the latter half of the 1950s. Although the firm never quite managed to catch up to the level of the leading European producers of military electronics, the 'buy national' arms procurement policy of the Government guaranteed the existence and even expansion of this AEG subsidiary. Somewhat better positioned in the

[12] See Immenga, U., 'Macht durch Staat', *Wirtschaftsdienst*, no. 12 (1988), pp. 603–5.

[13] *Stellungnahme zum Antrag der Daimler-Benz AG auf Ministererlaubnis* (Monopolkommission [Federal Monopoly Commission]: Bonn, 3 Aug. 1989), p. 14.

[14] See also chapter 11.

international league was AEG Naval Technology in Wedel, near Hamburg, a producer of naval electrical and electronics applications, mines and torpedoes.

At MTU, military production follows two different traditions. In one area of activity, MTU pursues a traditional specialization in fast-running, heavy diesel engines for naval and other applications, formerly sold under the label of Maybach. After taking control over the competing French firm Pilstick in 1990, MTU has gained a near-monopoly position in the West European market in this field. In another area, MTU is the preferred German partner in multinational co-production schemes for jet engines. While established in this demanding production sector for some time, MTU still does not have the capacity to develop jet engines on its own. For the time being, MTU is forced to rely on military and civil production co-operation programmes negotiated as part of larger work-sharing packages.

Dornier successfully resisted pressures to give up its independence until 1985, when disagreements within the Dornier family began to cripple company management. After the direct intervention of the Baden-Württemberg state Government, most members of the family agreed to sell a majority of their company shares to Daimler Benz. Dornier specialized in electronics research and development (R&D) and intensive application, with production focused on drones, small transport aircraft and space systems. Starting in the early 1980s, Dornier also pursued a systematic diversification into civilian high-technology markets, such as medical systems.

In many ways more an R&D enterprise than an arms production company, MBB had the largest and most diversified portfolio of all the companies coming under the Daimler Benz roof. About 25 per cent of its work-force were involved in R&D. Production sites were scattered all over the FRG, reflecting the successive mergers of aircraft manufacturers.

Daimler Benz itself has a successful record of supplying a broad variety of military vehicles to the German armed forces as well as to those of clients world-wide. Daimler Benz military automobiles and trucks have a high degree of commonality with civilian vehicles, which strengthens their competitiveness. It also means that they are largely exempt from arms trade restrictions.[15] Military production lines at Daimler Benz hardly differ from civilian ones, which allows for a great degree of flexibility in production. For instance, during the economic crisis following the second oil price increase in 1979, Daimler Benz increased its output of military trucks. The reduced demand on the civilian market thus had less effect on Daimler Benz than on its competitors, some of which fared badly.

The decision of the Daimler Benz Board to pursue the take-over was based on an assessment of a number of factors influencing the company's future, including the creation of the European Community (EC) Single Market, a possible decline in the growth of markets for large passenger cars and trucks,

[15] In order to fall under export control as stipulated in the appendices of the Foreign Trade Law, a product must be especially constructed, not just adapted, for military purposes.

and the fear of increasing competition from East Asian producers. The fact that diversification into high technology meant an enlarged role in arms production was a side-effect that created additional opportunities to increase Daimler Benz's future economic position. Operationally, there are two ways to use military production to bolster the relative economic position as well as corporate revenues. One is to optimize the military application of dual-use technologies, particularly at the component level. The other is to strengthen the firm's position in the arms market proper. The vertical integration of a huge corporation such as Daimler Benz should allow the transfer of technology from military to civilian applications and vice versa once the barriers between the two sectors inherited with the acquired companies are removed. For Daimler Benz the most important bridge between the two sectors is probably electronics.[16] It is at the heart of future competitiveness in engine and automobile as well as aerospace and arms production. This wide spectrum of applications is intended to put Daimler Benz in a privileged position to detect, implement and diffuse new advances in electronics.

IV. Military–political consequences of the take-over

The federal Government's decision to sell MBB—motivated by the ideological aim of ending state ownership of industry as well as the financial goal of passing on the responsibility for the seemingly limitless expenditure for Airbus subsidies—meant the end of a unique programme of government-nurtured high-technology companies. As far as its majority of public owners were concerned, MBB had become uncontrollable. Since the Government deemed all of MBB's production to be of strategic importance to (*a*) the security of the FRG, (*b*) the maintenance of West European autonomy in arms technology and production and (*c*) the high-technology international competitiveness of German industries, orders were less a question of performance than of political leverage to maintain the image of strategic assets. Neither the military technologies, the Airbus programme—the 'European' challenge to the US monopoly in large civilian transport aircraft—nor MBB's high-technology space products lent themselves to effective cost control, as none of the products had to face a competitive market. Sales were almost guaranteed.

By handing MBB over to Daimler Benz the Government rid itself of one dilemma, symbolized by having public officials sitting on both sides of the negotiating table when placing procurement orders with MBB. However, in doing so it was confronted with another, namely having the largest industrial group in the country as practically the sole source for a major part of its military procurement needs, especially in the area of R&D.

[16] For general arguments about the importance of electronics in arms production, see Bittleston, M., *Cooperation or Competition? Defence Procurement Options for the 1990s*, Adelphi Paper no. 250 (International Institute for Strategic Studies/Brassey's: London, spring 1990), p. 71; Reichelstein, H. E., 'Strukturwandel der Rüstungsindustrie', *Europäische Wehrkunde/WWR*, no. 7 (1989), pp. 387–90.

Given the regional distribution of arms production in Germany, the impact of this change in approach is felt especially in the Munich area. The deliberate attempt to establish modern industries in Bavaria depended on the particular interaction of state and private initiative and capital with a military demand. Bavaria is an outstanding example of what might be called military-led regional development in high technology.[17] However, the successful sponsoring of high-technology industries in Bavaria through large military R&D contracts has become a relative disadvantage. The insecurity of future military contracts haunts the Bavarian production centres of the German aerospace industry, as civilian aircraft and space production is predominantly located elsewhere. With Airbus subsidies more or less guaranteed, and the future of military projects such as the European Fighter Aircraft (EFA) insecure, the prospects for the advanced-technology industries in the Munich area seem bleak.

In addition, Daimler Benz has made good its declaration that it would regard the take-over of MBB as just a step towards the further internationalization of its business. It has concluded with French companies a number of strategic agreements for the production of various types of weapon systems, including vehicles (with Panhard), long-range missiles (with Matra) and helicopters (with Aérospatiale). In the area of jet engines, the Daimler Benz management has signed a far-reaching co-operation agreement with Pratt & Whitney (USA). In a large number of high-technology areas it is committed to close co-operation with Japan's largest industrial group, Mitsubishi—an arrangement which drew world-wide public attention (see also chapter 4).[18]

Co-operation in helicopter production has resulted in the establishment of a new company, Eurocopter, owned jointly by MBB and the French firm Aérospatiale. Future helicopter development has been transferred to Aérospatiale. This is the first instance in which Daimler Benz has been able to use national restructuring as an intermediate step towards transnational restructuring. In the near future, Daimler Benz's co-operation in both military and civilian areas (in particular automobile electronics) with Matra (of which it already holds a minority share[19]) might evolve in a similar way.

Daimler Benz has become more or less identical to the German aerospace industry.[20] It also controls about one-half of the first-tier arms industry. While military production and civilian aerospace production are of relatively minor importance in terms of the total output of the company, with shares of less than

[17] For a systematic discussion of *Kriegswirtschaft* as a stimulus to industrial development, see Sapir, J., *L'economie mobilisé* (La Decouverte: Paris, 1990).

[18] See Wagstyl, S., Marsh, D., Betts, P. and Oram, R., 'A door to the west', *Financial Times*, 7 Mar. 1990, p. 18.

[19] Although Matra is a private company, the French Government considered the sale of shares a sensitive issue, and gave its approval only after deliberations behind close doors.

[20] Of the 86 000 employees of firms affiliated to the German Association of Aerospace and Space Industries (BDLI), about 60 000 work within the Daimler Benz group. See Jäger, U., Schwegler-Rohmeis, W. and Berger, W., *Rüstung ohne Grenzen* (Verein für Friedenspädagogik Tübingen: Tübingen, 1989), p. 71.

10 per cent and 5 per cent, respectively,[21] Daimler Benz has still become an important political player. Its management has committed itself to expand its newly acquired high-technology sectors, in which arms production accounts for about 50 per cent of total output. The company has therefore begun to seek political support for special treatment that would favour it in competitive markets. Due to the particular circumstances that brought the merger about, it must do this in a very careful and quiet way. Reuter and his colleagues have in interviews and statements repeatedly stressed that they welcome disarmament and plan to decrease significantly the share of arms production in the total turnover. Already before the take-over, Reuter publicly accepted that military procurement and R&D would shrink in the near future. However, he predicted that space technology would take its place as the spawning ground for new technology, and that microelectronics and communication technologies would win the 'peace dividend'.[22] In other words, Daimler Benz welcomes all types of government contracts, but shows, at least publicly, a marked preference for those with civilian labels. At the same time, the company's managers warn of the consequences for its arms producing work-force of over 30 000 should the EFA be cancelled or should German arms export regulations remain stricter than those of other countries within the EC.[23]

V. Repercussions for smaller arms producers

This concentration on the level of the largest arms producers has had follow-on effects throughout the arms industry in southern Germany (see also chapter 11). Smaller producers have formed alliances, merged or been bought up by larger producers (see appendices A and B). While such changes might have occurred even without the Daimler Benz–MBB take-over, it increased their speed, for at least two reasons. First, as the Federal Cartel Office predicted, Daimler Benz put pressure on component manufacturers and subcontractors to produce more efficiently and to decrease costs. Second, and possibly more important, the take-over highlighted the changed perception of the military sector on the part of federal and local government. Other producers have had to cope with a tougher environment. They have tried to adapt, by increasing their share in arms production, attempting conversion or diversifying into civilian markets.

Diehl, a company in northern Bavaria originally specializing in tank components, such as chains and ammunition, has since the late 1970s tried to increase its production in high-technology areas, especially electronics. At the time of the Daimler Benz–MBB take-over there were some discussions among arms producers in the ordnance field, in particular Rheinmetall, Wegmann and

[21] See Deutsche Aerospace (DASA), *Das Geschäftsjahr 1990* (DASA: Munich, 1991), pp. 10–14. See also appendix A.

[22] Interiew in *El País* (Madrid), 30 July 1989, p. 5.

[23] Deutsche Aerospace (note 21); compare the introduction by DASA Board of Directors Chairman Jürgen E. Schrempp with the subsequent chapters on individual production sectors.

Krauss-Maffei, to create a counterweight to the new giant by uniting under the roof of Diehl. However, when nothing came of these talks, Diehl strengthened its ties with MBB in the production of explosives, rockets and 'smart' weapons. It also acquired, one by one, all large tank repair facilities in the FRG, including, in late 1990, one formerly used by the East German Volksarmee. Diehl is also involved in the destruction of former Volksarmee ammunition.

While Diehl has sought to increase its share in a shrinking arms market, the major tank producer Krauss-Maffei has successfully diversified into civilian fields, such as train propulsion and machinery for the plastics industry. While still committed to tank production, investments in recent years were clearly biased towards civilian applications. This trend was probably reinforced by the new dominant shareholder, the Mannesmann Group, which bought a large package of former MBB shares from the Bavarian state Government.[24]

Rheinmetall, a leading producer of artillery and artillery ammunition, has also reduced the share of military production in its total output, but rather by buying up civilian firms than by conversion. It closed the military production lines at its original factory in Düsseldorf, firing 1500 of the work-force. At the same time, Rheinmetall purchased the military section of Krupp MAK, in Kiel, a modern automated plant for tank assembly, and transferred military production to Unterlüß, Lower Saxony, a site heavily subsidized as a local 'technology park'. Thus, Rheinmetall succeeded in both diversifying into civilian manufacture and modernizing its military production.

Siemens has also pursued a double strategy of expanding in both civilian and military markets, the latter in a distinctly transnational way by acquiring, in a joint bid with General Electric of Britain, large parts of the British electronics producer Plessey. Some observers saw this move as a reaction to Daimler Benz's take-over of MBB:

> Daimler Benz will represent a unique case of concentration at the national level in Europe and even in the world, because this single entity will occupy 40 per cent of the budget for defence equipment in Germany. As a reply, Siemens, which was predicted to withdraw from the defence sector, has doubled its involvement in Plessey by concentrating on military telecommunication and radars. It has thereby become one of the first European transnationals in defence electronics.[25]

In an opposite way, Heckler und Koch, a traditional producer of small arms from the Black Forest area, has also become transnationalized. The company was put up for sale in late 1990 when the German procurement authorities

[24] In 1984 MBB wanted to acquire Krauss-Maffei from the Flick Group, which at the time was interested in selling the company. When the Federal Cartel Office blocked the take-over, the Bavarian state Government, led by Prime Minister Franz Josef Strauß, stepped in to acquire 35 per cent of the Krauss-Maffei shares, allowing MBB to acquire 12.5 per cent. The Federal Cartel Office could not object to this arrangement on legal grounds, even though it meant that MBB was able to control the management of Krauss-Maffei.

[25] Verret, D., 'L'Europe de l'industrie de l'armement', *Défense nationale*, no. 5 (1991), p. 122 (authors' translation).

announced that it would not proceed with the development of a new automatic rifle, the G-11. Heckler und Koch had invested heavily in the G-11 on the assumption that the Government would buy the finished product, in keeping with past arrangements. However, the Government postponed the decision to field the G-11, and refused to provide further subsidies to the firm. In late 1990 it looked as if the French firm GIAT would buy the almost bankrupt company, but in late December 1990 GIAT's Board of Directors finally decided against a purchase. Instead, Royal Ordnance, a subsidiary of British Aerospace, stepped in to acquire Heckler und Koch.

VI. Conclusion

The analysis presented here suggests an agnostic view of Daimler Benz's take-over of MBB. Daimler Benz is certainly not unwilling to take in all the military orders it can get, and will continue to improve its competitiveness in the field of arms production, but is mostly interested in remaining or becoming a world-class competitor in many types of air- and land-based transport vehicles. Its future prospects for dictating the terms of trade with the German Government are not good, partly because of the shrinking importance of the military sector in general, and partly because of the process of internationalization of procurement and production that Daimler Benz itself has furthered through its expansion. On the other hand, the take-over has put Daimler Benz, and in its wake other German producers, in a good position to increase their shares in a shrinking, but more internationalized, European arms market.

11. Crisis and adaptation in northern Germany

Werner Voß

I. Introduction

Prior to the opening of the border between the two German states, comparisons were often made in the Federal Republic of Germany of the economic development of the northern and southern parts of the FRG, the so-called south–north gap. In spite of the perceived dichotomy, however, no scientific consensus could be reached as to the geographic division of the two regions. Different interests and subjective factors determined where the dividing line was drawn. For the purposes of this analysis, besides the four coastal states of Schleswig-Holstein, Hamburg, Niedersachsen and Bremen, the city of Kassel in the state of Hessen is included in 'northern Germany', as the term is used here.

The shipbuilding (section II) and military electronics industries (section III) are both concentrated in northern Germany. The inclusion of Kassel allows the development of tank and tank-part production (section IV) to be explored as well. Large arms production facilities are concentrated in Kassel, providing direct employment for 4000–5000 workers in 1990. This represents about 10 per cent of the city's metal industry work-force, or 7–8 per cent of the trade industry work-force.[1] In comparison, the national average is about 3–4 per cent. This shows Kassel to be a leading centre for arms production in the FRG.

II. Restructuring of the shipbuilding industry

Shipyards in the FRG and elsewhere in Western Europe have since the mid-1970s experienced a production crisis that has threatened continued operation. The main factors behind this development are reductions in total demand and the appearance of foreign competitors, especially in South-East Asia. The crisis was preceded by an immense boom in demand, especially in the oil tanker market. While construction capacities were greatly increased world-wide in the early 1970s, from 1975 to 1987 the total displacement of merchant ships completed was reduced from a total of 34 202 000 to 12 259 000 gross registered tonnes (naval construction excluded).

Towards the end of the 1980s, however, the industry showed signs of remarkable recovery, to a great extent the result of concerted government and

[1] Weber-Bensch, J. and Strutynski, P., 'Umstellung der Industrieproduktion auf neue Erzeugnisse—Beispiele für Kasseler Betriebe', *FPN Arbeitsforschung und Raumentwicklung*, MIK 04 (Reihe Metallindustrie in Kassel: Kassel, 1990), p. 15.

industry efforts to restructure production. In the FRG, shipyard managers and politicians have applied the following measures:

1. In the civilian sector, construction capacities have been geared towards increased specialization (with production of ice-breakers, ferries, cruise ships, gas transport ships and other custom-fitted vessels).

2. Federal and state governments have placed large subsidies and other financial resources at the disposal of the industry.

3. Through personnel reductions and the consolidation and closing of shipyards, the cost structures of various companies have been reorganized, often against heavy protests from employees. As a result, the number of employees in the West German shipbuilding industry has declined from about 80 000 in 1970 to about 33 000 in the late 1980s.

4. Some shipyards have received direct help, usually through the military, to keep production facilities occupied. For example, although it would have been cheaper to limit the largest West German warship programme in recent years, the F-122 frigate, to a single contractor, it was parcelled out to the five largest shipyards in order to keep all of them in business. In addition, restrictive national regulations on arms export were slackened or lifted as they applied to naval construction. Depending on the source, the FRG in recent years ranks fourth to sixth among weapon exporting countries in global sales. Warships account for approximately 70 per cent of this export.

Construction and repairs for the Navy accounted for about 20 per cent of the total man-hours for 1989 of the West German shipbuilding industry. Table 11.1 lists West German shipyards working under military contracts in 1990. These figures do not include naval contracts planned by the Ministry of Defence but not yet assigned; these are to be offered to West German firms in the near future. As the Independent European Programme Group (IEPG) has exempted warship construction (as distinct from outfitting) from regulations requiring IEPG-wide competitive bidding on arms contracts, according to a decision taken in February 1990, shipyards in the FRG are not seen as threatened by the growing internationalization of the arms market.[2] Federal contracts will dominate in the future. The ending of the German participation in the joint NATO NFR-90 frigate programme is a sign of continued federal support for national—as opposed to joint—programmes.

With the exception of Bremer Vulkan[3] and Husumer Werft, in late 1990 shipyards producing for the military had contracts running at least until the end of 1991. Companies that are extremely dependent on navy contracts, such as Blohm & Voss (with arms contracts accounting for 50 per cent of total

[2] See *Wehrtechnik*, vol. 22, no. 4 (Apr. 1990), p. 19.

[3] Bremer Vulkan reserves 20 per cent of production capacities for military production; as a result of a fall in demand only 3–5 per cent are currently used. This situation is temporary, however: Bremer Vulkan participates in the Tender 404 project and the F-124 frigate programme, successor to the F-122 programme. Four frigates are to be delivered to the German Navy in 1994–96. Production started in 1991 and will lead to a greater utilization of naval capacities.

earnings) and Lürssen (almost 100 per cent), have contracts extending to 1996 or further. The Howaldtswerke-Deutsche Werft shipyard in Kiel, which is also heavily dependent on military contracts (about 50 per cent in 1990) is fully occupied with civilian and military contracts until 1993. Because of the current favourable state of the merchant ship market, there are only isolated problems, if any, for the shipbuilding industry in former West Germany (shipyards in former East Germany face large problems).

Responding to stimulation in the world economy in general and, specifically, to economic opportunities arising out of the political developments in Eastern Europe, shipbuilding firms' civilian and non-maritime production (such as the construction of energy plants) may in coming years profit from growing demand. West German shipbuilding firms should therefore better be able to cope with future reductions in military production than they have been in the past. However, whether these shipyards can survive in the long term without military programmes remains to be seen. In order to become internationally competitive in merchant shipbuilding, most firms have first put the price-settlement process (cost-plus-price, early payment) into effect in the arms sector.[4] Without these cost-transfer possibilities, many shipyards would in the past not have remained competitive and would already have been closed down. Past efforts at diversification, to the extent that it has been attempted, have been less successful.[5] In spite of large reductions in turnover, only about 20 per cent was in areas outside naval shipbuilding in 1989.[6]

III. The restructuring of the military electronics industry in northern Germany

When Daimler Benz was allowed to take over Messerschmitt–Bölkow–Blohm (MBB) by special permit of the FRG Minister of Economics, Helmut Haussman (see chapter 10), the company management had to agree to the sale of all production interests connected with naval shipbuilding. New owners had to be found for the naval electronics production facility formerly owned by

[4] See Bontrup, H.-J., *Preisbildung bei Rüstungsgütern* (Verlag Wissenschaft und Politik: Cologne, 1986).

[5] Voß, W., 'Reduction in armaments production: lessons to be drawn from adjustment to structural change in the shipbuilding industry in the Federal Republic of Germany', Working Paper no. 12 of the Disarmament and Employment Programme (International Labour Office: Geneva, Nov. 1989).

[6] Joint Paper of the Verband für Schiffbau und Meerestechnik (Organization for Shipbuilding and Marine Technology) and the Industriegewerkschaft Metall (Metal Workers Union), no title, Hamburg, Sep. 1989, p. 9. The international climate and the possible reduction of military orders force companies to change military-oriented management policies and to plan for new production lines. Blohm & Voss, a shipyard which, in the division of new construction, for over a decade has been wholly occupied with military production, has announced that it will restart merchant shipbuilding. Abeking & Rasmussen, an old shipyard near Bremen with 620 employees, is seeking to establish itself in the polymer market. The aim is to transfer knowledge from naval technology to applications of wind energy. In 1989, 220 rotors were ordered by customers in the FRG. In a few years 80–100 employees are planned to be fully engaged in this sector. Diversification activities of Bremer Vulkan have been established as a holding company under the name of Vulkan Industries. Its task is to structure and control the activities of the dozen or so firms and divisions engaged in areas outside of shipbuilding, which currently employ about 1000 persons.

Table 11.1. Military orders placed with German shipyards for 1990–96

Company/ *Land*	Military order[a]				Purchasing country	Construction period	Final delivery
	Ship type	Class	Number	Comment			
Abeking & Rasmussen	Minesweeper	343	3	. .	FRG	. .	Dec. 1990
Lower Saxony	Minehunter	332	3	. .	FRG	1991–95	1995
Blohm & Voss	Frigate	Vasco da Gama	1	. .	Portugal	. .	Nov. 1990
Hamburg	Frigate	. .	2	Module	Portugal	. .	. .
	Frigate	. .	1	. .	Greece		1992[b]
	Frigate	F-123[c]	1	. .	FRG	1989–96	1996
	Frigate	. .	10	. .	Australia/ New Zealand[d]	. .	By 2005
	Frigate	. .	2	. .	Turkey	. .	1995
Bremer Vulkan	Frigate	F-122	1	. .	FRG	. .	Mar. 1990
Bremen	Frigate	F-123[c]	1		FRG	1993–96	1996
	Tender	404	2	. .	FRG	1991–93	1993
Howaldtswerke–Deutsche Werft	Submarine	. .	1	Module	Brazil	. .	. .
Schleswig-Holstein	Submarine	. .	1	Module	Turkey	. .	. .
	Submarine	. .	2	Module	Brazil	. .	. .
	Submarine	. .	3	. .	Israel		
	Frigate	Vasco da Gama	2	. .	Portugal	. .	1991
	Submarine	. .	3	Module	South Korea	. .	. .
	Submarine	. .	2	Module	Turkey	. .	. .
	Submarine	. .	3	Module	South Korea	. .	. .
	Frigate	F-123[c]	1	. .	FRG	1993–95	1995
	Submarine	. .	1	Rebuilding	Greece	. .	. .
	Submarine	. .	3	Rebuilding	Greece	. .	. .
	Submarine	. .	2	With TNSW	Unspecified	. .	. .
	Submarine	206-A	6	Rebuilding	FRG	. .	1991

Husumer Schiffswerft *Schleswig-Holstein*	Tug	725	3	..	FRG	..	1991
Kröger Werft Schacht–Audorf *Schleswig-Holstein*	Tug	745	3	..	FRG	..	1990
	Minesweeper	343	3	..	FRG	..	Dec. 1990
	Minehunter	332	3	..	FRG	1991–95	1995
Fr. Lürssen Werft *Bremen*	Corvette	FPB 62	2	..	Unspecified	..	1990
	Minesweeper	343	4	..	FRG	..	Aug. 1990
	Minehunter	332	4	..	FRG	1991–95	1995
Neue Flensburger Schiffbau-gesellschaft *Schleswig-Holstein*	Tender	404	2	..	FRG	1991–93	1993
Norderwerft *Hamburg*	Frigate	..	1	'Augsburg'; rebuilding as training ship	FRG	..	1990
Thyssen Nordseewerke (TNSW) *Lower Saxony*	Frigate	F-122	1	..	FRG	..	1990
	Submarine	..	6	..	Norway	1987–92	1992
	Frigate	F-123[b]	1	..	FRG	..	June 1995
	Submarine	206 A	6	Rebuilding	FRG	..	1991

[a] The table does not give figures for naval repairs.

[b] Three additional frigates are being constructed at the Greek Hellenic Shipyards, Greece; final delivery is set for 1997.

[c] Blohm & Voss is in overall charge of the F-123 consortium. One frigate is to be constructed at the Hamburg yard between 1991 and 1994; Bremer Vulkan, Howaldtswerke-Deutsche Werft and Thyssen Nordseewerke will construct one frigate each.

[d] Construction in Australia under licence and monitoring from Blohm & Voss

Sources: *Hansa*, vol. 127, no. 3 (1990) pp. 153–58; 'Abteilung "Schiffe, Schiffsgerät und Unterwasserwaffen" (SG)', *Wehrtechnik*, vol. 22, no. 1 (Jan. 1990), pp. 34–40.

Daimler Benz via its subsidiary company AEG and for the MBB divisions working on naval projects. The two entities, renamed Deutsche Marinetechnik GmbH (DMT) and Marine- and Sondertechnik GmbH (MSG), were made independent. At their production facilities in Hamburg and Bremen the two companies had close to 3000 employees and were expected to produce a military turnover of about 80 per cent of total turnover.[7] Although the firms are involved in some non-naval projects, naval contracts—more specifically the management of minesweeper projects and the production of maritime electronics—predominate.

A number of potential buyers showed interest in acquiring one or both of the two companies. The two state governments of Bremen and Hamburg, which as major MBB shareholders had to agree to Daimler Benz's take-over, made it clear that they preferred a north German solution. In the end Bremer Vulkan, with the backing of the two state governments, bought both DMT and MSG. There is an industrial and technological rationale behind Bremer Vulkan's purchase: the company wants to secure and expand its position as a leader in civilian and military shipbuilding. Concomitantly, technological leads are providing spin-offs in non-naval areas as well. For instance, electronics used in outfitting civilian as well as military crafts has been targeted for use in environmental protection technologies.

Whether this strategy will function is questionable, given the expected drop in the current high level of military funding. In order to expand in the increasingly competitive civilian market of the European Community, some electronics companies are going in other directions, away from arms-related production.

An obvious example is the Philips Corporation in the Netherlands, which disposed of its military production capacities world-wide in a very short time. In early 1990 Philips also sold its north German arms production facilities, which had been combined in 1985 to form the division of systems and special technology.[8] In 1989 both production divisions in Bremen and Kiel, each with about 500 employees, had a turnover of about 106 million D-Marks. In the beginning of 1990 Philips turned this division into an independent company named Deutsche System-Technik GmbH (DST); two previous Philips managers subsequently acquired 50 per cent each of the capital. The company's continued operation seems currently assured by existing long-term contracts.[9]

In spite of the optimism of the new owners of these companies, it is questionable whether the growing development and production costs for military equipment, under pressure from shrinking defence budgets and an increasingly competitive international market, will allow the companies to

[7] *Weser Kurier*, 28 Mar. 1990; *Weser Kurier*, 4 Apr. 1990.

[8] Earlier Philips sold its arms production facilities in Sweden to Bofors, and a majority of its shares in Hollandse Signaalapparaten B.B. and the Belgian MBLE to French Thomson-CSF. Thomson-CSF also acquired Philips' shares in French TRT. See *Financial Times*, 23 May 1989; *Die Welt*, 8 Aug. 1989.

[9] *Süddeutsche Zeitung*, 10–11 Mar. 1990. At the Bremen facility, formerly Elektro Spezial, the main focus is on optronics, communications and simulation; the Kiel facility, formerly Radarleit GmbH, specializes in ship electronics, radar and computer hard- and software.

remain viable commercial producers of high-technology products, especially of arms equipment.

Employee groups, among others, have presented conversion to civilian production as an alternative course, but there is substantial inertia in the industry. Timm Meyer, responsible for military–technical industry at the Federal Organization of German Industry (*Bundesverband der Deutschen Industrie*, BDI), said in an interview in May 1990:

> Conversion does not necessarily mean that, at the moment, there is a divergence from the civil areas, but rather that there are also conversions in the military sector. As you know, the consequences of disarmament negotiations will be that new disarmament technologies will have to be developed. It is possible that new technological needs will arise from requests for a reduction in fighting forces, as this will present different technological demands than those of the present.[10]

For the Bremen firm Krupp Atlas Elektronik (KAE), the third largest military electronics producer in northern Germany with about 4400 workers, these new developments will not require any large conversions. KAE has since the mid-1960s dedicated much of its effort to the development of electronic simulation technology, which could play a large military role in the future. By replacing manœuvres and other off-base exercises with on-base simulators for specific training tasks, problems with the public and costs can be reduced. Such technology also allows for greater reliance on simulation testing of new military equipment, thereby reducing research and development costs. KAE also develops and manufactures sonar installations for submarines and firing installations for tanks. The firing installations for the Leopard 2 tank are manufactured at the KAE facility in Bremen.

IV. The tank and tank-part industry in northern Germany

The increase in tank production in the FRG after World War II is closely linked to the Leopard family of tanks, which constitutes the core of the West German Army. The latest version, the Leopard 2, is produced jointly by Krauss-Maffei in Munich and Krupp MaK GmbH in Kiel. Along with KAE, the Blohm & Voss shipyard (Hamburg; tank hulls) and Wegmann & Co. (Kassel; tank turrets) are important suppliers in northern Germany. In addition, Thyssen Henschel manufactures tanks and tank parts. Tanks are also manufactured in a facility in Flensburg which Diehl, one of the largest arms production companies in the FRG, took over from the former Harmstorff Group in 1986.

With the delivery of the eighth batch of Leopard 2 tanks, production of this model at Krupp MaK was completed. The military share of the firm's total production nevertheless remains a strong 50 per cent. Pionier armoured engineering vehicles are modernized and many smaller military projects are being

[10] Quoted from Mayr, G. and Beyer, G., 'Scharfe Schwerter, stumpfe Pflugscharen?', radio manuscript, Westdeutscher Rundfunk, May 1990 (in German).

completed. Preparations are also being made for serial production of a new type of armoured recovery vehicle as well as of tank howitzers, but it is questionable if these plans can be realized. Should there be delays or cancellations in these programmes, MaK will face serious difficulties in maintaining production.

In view of these difficulties, the management of Krupp decided to sell the military division of MaK and to tie the civilian production areas more closely to the parent company. In 1990 the military production section was transferred to a new company, MaK System GmbH, a majority share of which was bought by the country's largest producer of cannons, Rheinmetall, leaving Krupp with a minority share. This sale, which emphasized Rheinmetall's role as a military producer, might make it more difficult to stabilize employment at MaK System GmbH. Before the transfer of the military division the 600 employees might have been transferred to the company's civilian production areas, specializing in motor production and rail-traffic technology. Now this is blocked, although with the creation of the EC Single Market and the attention given to infrastructural needs in the former German Democratic Republic there seems to be good chances to increase civilian lines of production. However, Krupp faces strong competition, especially in the market for rail-traffic technology which, with 16–18 producers of locomotives in Europe, is burdened by large overcapacities.

The second largest tank producing company in northern Germany, Thyssen Henschel in Kassel, is facing a similar problem.[11] It also combines tank and locomotive production, but in this case locomotive production was strengthened and separated by a merger with the rail-traffic division of Asea Brown Boveri, effective from 1 October 1990.

Military production clearly dominates the output of Wegmann & Co., the second largest arms company in Kassel, accounting for approximately 85 per cent of total turnover. The main products are tank turrets, tank hulls and gun carriages. Wegmann & Co. is currently improving some 1100 Leopard 1 tanks. This has allegedly assured 'work for a number of years'.[12]

V. Summary

Analysis of the three branches of the German arms industry that are significantly represented in northern Germany (warship production, military electronics and tank production) has shown very different tendencies.

[11] Thyssen Henschel has four main divisions at two locations. Arms production and traffic technologies are concentrated at the Mittelfeld factory. Most employees (more than 1000) work in traffic technology. About 800 are employed in arms production. Currently the main military project is the upgrading of the Marder infantry vehicle, to be completed in 1993. In 1990 the NBC reconnaissance vehicle Fuchs was in production for the West German Army. See *Wehrtechnik*, vol. 20, no. 5 (May 1988), p. 110; *Wehrtechnik*, vol. 22, no. 2 (Feb. 1990), p. 58.

[12] Weber-Bensch, J. and Strutynski, P., 'Skizzen der fünf größten Metallbetriebe im Kasseler Raum', *FPN Arbeitsforschung und Raumentwicklung*, MIK 02 (Reihe Metallindustrie in Kassel: Kassel, 1989), p. 96.

The warship as well as the merchant ship industries are, in the short term, occupied to capacity, and no great changes are to be expected in the near future. This is reflected in IEPG calculations, in which military shipbuilding has already been removed from the Group's consolidation process. Measures to diversify production are already being taken by some shipyards. However, it is questionable whether companies that build warships will be able to survive a reduction in demand in the mid-term. Past experience with such diversification efforts shows little promise. Production potential and management efforts were insufficient to effect successful conversion.

Several companies in the microelectronics industry in northern Germany have changed owners in 1990. With few exceptions, this has not been a process whereby larger companies already in the market have annexed production capacities, but rather of a separation of military production facilities and plants from civilian ones. The new ownership structure is characterized by an increased number of small, independent companies. The future will show whether companies that for decades have depended for their survival on the protectionist policies of the federal Government will be able to remain in the market. Faced by cost increases and over-production capacities in Western Europe, this seems possible for only a few firms. Competition from traditionally civilian-oriented companies, resulting from the military application of innovations in civilian technologies, may also be on the increase. In the mid-term, plant closures and changes of ownership are therefore to be expected in the military electronics industry in northern Germany.

The future of the tank industry in northern Germany also looks unsure. The usefulness of tanks has already come into question as a result of the radical changes in recent years in East–West relations and the accompanying discussion of military doctrines, and this tendency will likely be strengthened in the years to come. This tendency increases the pressure to concentrate manufacturing capacities on a West European level. Whether the West German companies in tank production will survive is unsure. Since the largest West German producers, Krauss Maffei and MaK, have already found new owners, Thyssen Henschel is a promising candidate for take-over. With the possible withdrawal of British military forces from the FRG, however, the future of the British tank industry is even more uncertain. This could allow for a division of labour between the two allies, with the tank production industry consolidated in the FRG, while in the United Kingdom emphasis could be placed on other areas of military production.

12. The slow restructuring of the Italian arms industry

Giulio Perani and Mario Pianta

I. Introduction

Italy's arms industry has lagged behind in the extensive restructuring of the West European arms industries that began at the end of the 1980s. Up until 1990, the continuing growth of domestic procurement more than offset the rapid fall of Italian arms exports. This has postponed the need for rationalization and concentration of an inefficient industrial structure, which though largely controlled by three major industrial groups is fragmented in small production units with little international orientation.

This chapter gives a profile of the Italian arms industry, assessing its past performance and scenarios of adjustment to reduced arms spending. The strategy of the largest firms and the emerging pattern of restructuring are examined, pointing out the weakness of Italian industry in the European context.

II. The profile of the Italian arms industry

This section draws on material from a recent detailed study of the Italian arms industry that considers the top 50 military firms and subsidiaries, identified on the basis of the importance of their production of specific military equipment and materials.[1] Data on sales and employment, types of production and ownership have for each company been drawn from statistical yearbooks, company budgets, press information and trade union sources.

Ownership and production

Two-thirds of the top 50 military firms and subsidiaries are controlled by three major industrial groups: the state-owned holdings IRI and EFIM, and the private group FIAT.

IRI aerospace and electronics activity is controlled by the Finmeccanica group (with Aeritalia and Selenia), and shipbuilding is concentrated in Fincantieri. Other activities range from engineering to telecommunications.

Finmeccanica has emerged as the leading group in the Italian arms production sector. Aeritalia is the largest firm, producing civilian and military aircraft,

[1] Pianta, M. and Perani, G., *L'industria militare in Italia* (Edizioni Associate: Rome, 1990).

and is the Italian partner in the international co-production of the Tornado and the European Fighter Aircraft (EFA). Aeritalia has some control over most other military aircraft firms, including Alfa Romeo Avio, Officine Aeronavali, Elmer, Partenavia, Magnaghi, Aermacchi and Piaggio. Selenia is the major military electronics firm of Italy, producing radars and surface-to-air missiles, as well as civilian systems such as air-traffic controls. Aeritalia and Selenia were in 1991 merged into a new company, Alenia, integrating the aerospace and electronics activities of Finmeccanica.

The four Fincantieri shipyards build aircraft-carriers, cruisers, destroyers, frigates and submarines; this group has close production and marketing links to consortia of other firms producing naval electronic and weapon systems.

The second state-owned industrial holding is EFIM, whose artillery and electronics production is grouped in the Finbreda sub-holding, while aircraft production is dominated by the Agusta group. Finbreda's Oto Melara is the major producer of armoured vehicles and naval and field artillery, and is the largest Italian military firm with no civilian production. Also involved in artillery production is Breda Meccanica Bresciana, while Finbreda's main electronics companies are Officine Galileo and Sma, specialized in optronics and radar systems.

FIAT, the largest private Italian industrial group, has arms activities ranging from aero-engines (FIAT Aviazione) and heavy vehicles (IVECO) to chemicals (Snia Bpd) and components (Gilardini). FIAT Aviazione, which produces under foreign licences, is the Italian partner in the European co-production of the engines for the Tornado and the EFA. Snia Bpd, Gilardini and other FIAT subsidiaries have gone through an extensive reorganization, started in 1988, and aimed at trimming and regrouping FIAT's diverse military production.

Other Italian arms producers include the Intermarine shipyard, owned by Montedison; Aermacchi (aircraft) and Elettronica (electronics), both with a minority share of Finmeccanica; and the small arms producer Beretta.

Foreign firms with a significant presence in Italy include specialized military producers, such as Oerlikon in ground armaments, and large diversified companies (Alcatel, GEC-Marconi, Ericsson and United Technologies) active in telecommunications, electronics and aircraft components.

Almost all firms and subsidiaries have some civilian activity alongside military production, and the degree of diversification at the level of industrial groups is substantial.

The regional distribution of the top 50 military firms shows that 60 per cent of all plants and employment are in northern Italy, especially in Lombardy (25 per cent) and in Piedmont and Liguria (12–13 per cent). Latium and Campania (13–15 per cent) are the only central and southern regions with substantial military employment. The spatial concentration of military production is evident in the Italian case; Lombardy has the widest array of arms production, including aircraft, engines, light arms and electronics. Most aircraft production takes place in Lombardy (Varese) and in Piedmont (Turin). Liguria, with Genoa and La Spezia, is the centre of naval and mechanical production. The major electronics firms are concentrated in Rome.

The size of military production

For each of the top 50 military firms and subsidiaries, an estimate of the share of military production is defined, using company information, press articles and trade unions sources. Table 12.1 provides 1988 data for employment, sales and value added related to military production for the top 50 military firms.

In 1988 these military firms had a total employment in military production of about 55 000 people, military sales estimated at $6.3 billion, and a value added of about $2.5 billion. These data represent about 40 per cent of the total activity of the 50 firms.[2]

Compared to the total manufacturing industry in Italy, military production of the top 50 firms accounted for about 1.3 per cent of all employees and of total manufacturing value added in 1988. Furthermore, estimates of both military production and total activity of the 50 companies for the 1980s showed a declining importance in the overall Italian manufacturing industry. Estimates of military sales and value added were stagnant in real terms in 1983–88, while total sales and value added of the 50 firms increased over the same period by 17 per cent and 12 per cent, respectively, while value added in total manufacturing industry grew by 21 per cent in real terms.

The list of military firms in table 12.1 shows that the largest military employment in 1988 was in Aeritalia (8850 persons), followed by two other IRI companies, Fincantieri (4650) and Selenia (4550). In FIAT Aviazione and Agusta, military activities employed 3900 and 3100 persons, respectively; two highly specialized firms follow: Aermacchi (2500) and Oto Melara (2300). Companies with fewer than 2000 employees in military production are Snia Bpd (1950), FIAT IVECO (1550), Elettronica (1250), Marconi Italiana (1250), Siai Marchetti, Elmer and Contraves.

More than 21 000 people (almost 40 per cent of all persons employed in military production) are employed by IRI , while EFIM and FIAT employ about 10 000 persons each. Together these three groups account for 80 per cent of all employees involved in the military activities of the top 50 firms.

Looking at military sales, the largest firms are Aeritalia (with sales worth $880 million in 1988), FIAT Aviazione, Oto Melara and Agusta, followed by Selenia, Fincantieri and FIAT IVECO, all with 1988 sales greater than $200 million.

Also in terms of value added, Aeritalia is the largest firm (about $340 million in 1988); Oto Melara, Selenia, Agusta, FIAT Aviazione, Aermacchi and Marconi Italiana follow with values greater than $100 million.

[2] These data confirm, to a large extent, the results presented in Battistelli, F., *Armi: nuovo modello di sviluppo?* (Einaudi: Turin, 1982). Estimates of total employment related to military production, including the top 50 firms, and the military work carried out in a few hundred smaller companies and major subcontractors, indicate a total of about 80 000 people.

With sales and employment of this size, Italy's arms production is a marginal part of the country's manufacturing industry and is much smaller than the arms industry in other major Western European countries, such as the Federal Republic of Germany, France and the United Kingdom.

The performance of the arms industry

These data show a generally low ratio of value added to total sales. This is due to several factors, including the fragmented structure of the industry, the small size of most firms and subsidiaries, the practice of decentralizing part of the production in several companies and, most importantly, the need to import key sub-systems, components and technology. The ratio of value added to sales has shown some growth in recent years; with falling orders, many firms—Oto Melara is the most extreme case—have increased the share of work performed in-house, reducing the degree of subcontracting to smaller independent firms.

The value added per employee in the largest military firms is higher than the average of Italian manufacturing industry, due to the larger number of white collar workers and engineers. However, the rates of change of value added per employee, a better indicator of productivity growth, show a poorer performance of military production compared to Italian manufacturing industry.

An indicator of the firms' financial situation is the level of their debt. In 1988 almost all of the major companies were heavily indebted: Fincantieri for over $760 million, Agusta for $630 million, and Aeritalia, Selenia and Oto Melara for $400–440 million each. For several companies, total debt is twice the size of their annual value added.[3] This indicates that the growth of most Italian military firms in the 1980s was based on extensive borrowing; in the present decline of military orders this financial exposure increases the weakness of many firms.

Relative to their total (military and civilian) activities and to the overall performance of Italian manufacturing industry, the performance of military production in the top 50 firms has been poor during the 1980s.

This result is particularly evident in a comparison of the index of sales at constant prices from 1983 (=100) to 1988 for military production and for Italy's major manufacturing branches. The estimate of military sales of the top 50 firms reached an index of 104.5 in 1988; their total sales (military and civilian) reached an index of 117.2, a value greater than the index for the branches of mechanical machinery and other transportation equipment. Other manufacturing sectors have shown a much faster growth: chemicals reached an index of 125; electric material, 135; motor vehicles, 141; and office machinery, 163.

The conclusion which can be drawn from the available evidence is that the Italian arms industry is not only a very small part of the national manufacturing system, but is also an industry showing slower growth, lower productivity increases, and less efficiency and competitiveness than civilian sectors.

[3] Pianta and Perani (note 1). Data on individual company profits are not reported here, as they are less reliable as an indicator of a firm's performance.

Table 12.1. Military share of production, employment, sales and value added of the top 50 Italian military firms, 1988

Sales and value added figures are in US$ m.

Company	Military share of production (per cent)	Military employment (thou.)	Military sales	Military value added
IRI				
Aeritalia	*60–65*	8.85	882.6	343.8
Fincantieri	*20–25*	4.65	310.8	81.4
Selenia	*65–70*	4.55	379.9	203.4
Alfa Romeo Avio	*50–55*	0.79	82.9	36.1
Vitroselenia	*75–80*	0.59	67.5	28.8
Italtel SIT	*6–8*	0.65	75.6	32.2
Elsag	*30–35*	0.60	101.3	31.5
Officine Aeronavali Venezia	*65–70*	0.57	36.1	16.9
Ansaldo Sistemi Industriali	*5–10*	0.12	16.1	4.4
Total IRI	..	**21.37**	**1 953.0**	**778.0**
EFIM				
Agusta	*70–75*	3.10	491.2	175.0
Oto Melara	*99*	2.30	533.0	214.9
Siai Marchetti	*60–65*	1.08	69.1	36.8
Officine Galileo	*70–75*	0.93	90.6	49.1
Elicotteri Meridionali	*70–75*	0.83	55.3	27.6
Breda Meccanica Bresciana	*95*	0.70	119.0	45.3
Sma	*95*	0.53	51.4	30.3
Omi	*85–90*	0.43	38.8	14.6
Iam	*30–35*	0.31	16.1	9.2
Caproni Vizzola	*65–70*	0.25	13.0	9.2
Breda Nardi	*70–75*	0.17	9.7	5.4
Total EFIM	..	**10.63**	**1 487.0**	**617.0**
FIAT				
FIAT Aviazione	*90–85*	3.90	660.1	158.9
Snia Bpd	*55–60*	1.95	155.0	47.6
FIAT IVECO	*5–10*	1.55	268.6	72.9
Gilardini	*10–15*	0.60	69.1	24.6
Simmel Difesa	*95*	0.55	52.2	23.0
Telettra	*7–9*	0.45	43.7	33.0
Borletti FB	*95–100*	0.38	29.9	13.0
IVECO Aifo	*80–85*	0.37	179.6	20.0
Sepa	*50–55*	0.27	23.0	12.7
Sipe Nobel	*50–55*	0.19	23.4	11.4
Misar	*90–95*	0.11	13.8	6.1
Astra Veicoli Industriali	*10*	0.07	11.5	3.8
Total FIAT	..	**10.39**	**1 530.0**	**427.0**
Montedison				
Intermarine	*100*	0.49	28.2	7.2
Arturo Junghans	*100*	0.27	16.7	8.9
Total Montedison	..	**0.76**	**45.0**	**16.0**

Company	Military share of production (per cent)	Military employment (thou.)	Military sales	Military value added
Other Italian				
Aermacchi	*90–95*	2.50	182.7	119.0
Rinaldo Piaggio	*65–70*	1.30	109.0	50.7
Elettronica	*95*	1.25	131.2	77.5
Europa Metalli LMI	*20–25*	0.44	115.1	28.4
Riva Calzoni	*30–35*	0.33	33.0	15.7
Beretta	*15–20*	0.18	17.7	6.9
Larimart	*60–65*	0.09	10.0	4.2
Rodriguez	*10*	0.03	2.3	1.2
Total other	..	**6.12**	**601.0**	**304.0**
Foreign-owned				
Marconi Italiana	*50–55*	1.25	145.8	110.5
Elmer	*100*	1.05	119.6	..
Contraves Italiana	*99*	1.03	194.9	62.9
Microtecnica	*70–75*	0.80	76.8	39.1
Fiar	*55–60*	0.70	63.7	30.7
Oerlikon Italiana	*85–90*	0.67	93.6	33.0
Industrie Face Standard	*7–9*	0.37	36.5	15.0
Total foreign	..	**5.87**	**731.0**	**291.0**
Total	..	**55.14**	**6 347.0**	**2 434.0**

Figures may not add up to totals due to rounding.

Source: Pianta, M. and Perani, G., *L'industria militare in Italia* (Edizioni Associate: Rome, 1990).

The demand for Italian arms production

The main market for Italian arms production is the Ministry of Defence. Data from the defence budget suggest that total arms-related expenditures—acquisition, maintenance, research and development (R&D), etc.—increased from $2.3 billion in 1983 to $5.1 billion in the budget appropriations for 1990, with a 53 per cent increase in real terms in 1983–88.[4] The procurement market is highly protected; from the available data on military contracts assigned in 1985–88, only 7.4 per cent went to foreign firms, even though Italian firms rely on imports of key subsystems for their arms programmes.

This rapid growth of national procurement has offset the dramatic fall of arms exports; according to SIPRI data, Italian arms exports reached a peak of $973 million in 1983 and fell to $438 million in 1988 and to $149 million in 1989. Most sales went to Third World countries, where arms purchases have

[4] A series of comparative studies on arms procurement in West European countries, including Italy, can be found in De Cecco, M. and Pianta, M. (eds), *La spesa per armamenti in Europa* (forthcoming).

rapidly declined and several regional conflicts have ended. In other Western countries, Italian arms exports have been unable to find new markets because of the low technological levels of Italian weapon systems and the strong competition from other countries. As a share of total Italian arms production, exports have fallen from about one-third in 1983 to about 5 per cent in 1987,[5] and even the growing political instability in many areas, such as Eastern Europe and the Middle East, is not likely to provide new significant demand for Italian arms exports.

Only the rapid growth of new national arms projects (including the Tornado aircraft, the AM-X fighter aircraft, the MB-339 trainer aircraft, the *Garibaldi* aircraft-carrier, the wheeled–armoured Centauro vehicle, the A-129 combat helicopter and the Ariete battle tank) has allowed Italian arms firms to maintain their previous size and activity, so far avoiding major job losses. However, with increasing pressure for cuts in the defence budget and restructuring moves by many firms, the outlook for military production and employment appears bleak.

Scenarios of adjustment

The current decline in military spending and export will have a major effect on the profile of the Italian arms industry in the mid-1990s. The possible impact of reduced demand can be estimated using scenarios similar to those presented in chapter 2.[6] The first scenario, assuming a 3 per cent annual reduction in procurement spending in 1990–95, the same arms exports as in 1985–89, and a 2 per cent productivity increase, would for the top 50 Italian firms mean a fall in employment from 55 000 to less than 42 000 by 1995, with 13 200 jobs lost (out of a total European reduction of 340 000 jobs).

The second scenario, assuming a 5 per cent annual cut in procurement budgets and a 3 per cent loss in arms exports, with the same 2 per cent annual productivity increase, would for the top 50 Italian firms lead to a loss of 18 000 jobs (out of a total European reduction of 475 000 jobs).

While these data are rough estimates and need to be considered with caution, they show the serious industrial and employment effects of the likely reduction of arms production in Italy and in Europe as a whole. In order to prevent such losses of productive capacity and employment, an adequate conversion strategy is needed for the Italian (and the European) industry.

[5] For the method used for this estimate, see Pianta and Perani (note 1), chapter 5. In 1990, legislation was for the first time passed in Italy restricting and regulating the arms trade.

[6] See pp. 33–35. A more detailed analysis of the scenarios of adjustment for Italy's military industry can be found in Pianta, M. and Castagnola, A., *La riconversione dell'industria militare. Le strategie per disarmare l'economia* (Edizioni Cultura della Pace: San Domenico di Fiesole, 1990). However, the impact of reduced arms spending on Italian industry is likely to be softer than one could expect because of the large backlog of previous orders; the defence budget contains as yet unspent allocations to a sum of approximately $5.4 billion for arms programmes that have piled up in the years of fast procurement growth; this is equivalent to all arms-related expenditure for one year.

III. The weakness of the Italian arms industry

The structure and the performance of Italy's arms industry suggest a serious weakness, especially when compared to those of other West European countries. In this section a few key aspects of the condition of the Italian arms industry are briefly examined.

Concentration and fragmentation

The pattern of ownership of the Italian arms industry implies a highly concentrated structure, with three large groups accounting for four-fifths of the employment in the top 50 companies. However, looking at individual companies and their production units the fragmentation of the industry becomes apparent, with a low average number of employees involved in military production and the activities of firms and subsidiaries scattered across many different product areas, often with little integration even within the three largest industrial groups.

A strong polarization emerges between, on the one hand, a dozen large companies which concentrate on major arms production projects and Ministry of Defence contracts and, on the other hand, a few hundred small firms that depend on prime contractors and maintain a more diversified activity.

Although the four largest companies account for 40 per cent of military employment and sales, even the largest firms have difficulty in acquiring a capacity to develop entire weapon systems. Most firms therefore do not move beyond a 'niche' position in the national market. Such a combination of concentration and fragmentation has also shielded most companies from any effective competitive pressure, from either domestic or foreign firms.

Technological weakness

The R&D effort of Italian arms producers is rather low, and the only available data for government funding of military research suggest a value of about $690 million in 1988. After widespread acquisition of foreign production licences in the 1970s and early 1980s, Italian military firms did not succeed in developing 'indigenous' technology, in spite of the massive support of national arms acquisition programmes (such as the AM-X fighter aircraft, the EH-101 helicopter and the Catrin command and control system, funded by a special law). Even less successful was the attempt to win markets in Western countries.

The technological levels of the Italian arms industry can be compared with the position of other major countries using as an indicator the number of patents granted in military-related fields. Patent data are widely used in the analysis of national strengths and weaknesses in technology; to the usual caveats in the evaluation of the results, an additional caution is needed in the case of the arms industry, where secrecy is particularly important to preserve a technological advantage. However, in a wide range of military-related fields, patent protection remains a useful tool for appropriating returns from innovative activities, and this is confirmed also by the extensive international licensing of patents to foreign military firms.

Data from a study of the technological specialization of advanced countries[7] allow a comparison of the patenting activities of major countries in the US Patent Office, in the military-related classes of the Standard Industrial Classification (SIC).

The total number of patents granted by the US Patent Office and patent citations received from later patents (an indicator of the impact a patented invention has) between 1975 and 1988 have been considered, for the SIC classes of guided missiles, shipbuilding, aircraft and components, and ordnance (the four classes where military production is most relevant).

Italian military production shows the lowest technological level among European countries. Italy shows a small positive specialization only in ordnance for patent counts and in aircraft and parts for patent citations. France has indexes greater than one in all four classes; the FRG is specialized in all fields with the exception of shipbuilding; the UK has a strength in all classes but ordnance; and Sweden and the USA are specialized in all fields but aircraft. All data show a consistent picture for both patent counts and patent citations.

While this evidence must be considered with some caution, the technological weakness of Italian industry in military-related fields is clear. These results support the findings on the relatively poor performance of Italy's top 50 military firms, and suggest that even the dozen largest firms, the only firms able to operate on a European scale, are poorly equipped for facing the wave of international agreements that will affect West European military industries.

However, this pattern must be set in the context of the growing internationalization of industrial production in all sectors. As the degree of specialization has been increasing in most industries for most countries,[8] it is no surprise that it should also increase for military production. As Italy has traditionally shown a relative weakness in this field, its innovative activities have tended to concentrate in other industrial sectors, from traditional industries to machine tools, where Italian technological specialization has increased. Conversely, countries such as France and the USA have increased their specialization in military-related fields.

This is a pattern which will increasingly affect the restructuring of the arms industry in the Europe of 1993, when the pressure for growing specialization and competitiveness will be much greater even in military production. In this perspective the policy of maintaining, through a growing arms procurement budget, a subsidized and protected national industry is likely to be unsustainable in the longer term, as the high costs involved would continue to increase,

[7] Archibugi, D. and Pianta, M., *The Technological Specialization of Advanced Countries*, CNR-ISRDS Report (Consiglio Nazionale delle Ricerche: Rome, 1990).

[8] See Archibugi and Pianta (note 7).

and as it would require that industrial resources be diverted from areas of the national economy that enjoy greater specialization and competitiveness.

IV. The slow restructuring

The Italian arms industry is marked by a paradox. Along with France, Italy is the West European country with the most extensive presence of state-owned military firms. Still, far from being a more effective industrial policy tool, the presence of large state holdings with close links to different government parties has increased the institutional rigidity of the sector. Such a structure contributes to the fragmentation of the industry, leads to the duplication of production capacities and makes it more difficult to adjust to change. While in the private sector some restructuring has taken place, especially within the FIAT group, the state holdings have found it easier to pressure the Government for more contracts than to adjust to a falling demand.

This explains the lack of dramatic job losses in the late 1980s; however, this delayed restructuring makes the outlook for the future worse, as the inertia of the present structure prevents the development and implementation of appropriate industrial strategies. The institutional rigidity of the system linking political and military power, government parties and state industry remains the major problem for adjusting to a reality of reduced military threats, spending and production. Bearing this in mind, four major restructuring strategies can be identified in the Italian arms industry. These can be summarized as follows:

1. *Concentration and consolidation.* The standard response to an industry's declining demand is greater concentration, with many firms leaving the sector, others taken over by the stronger competitors, and a consolidation of production activities with lay-offs and streamlining. This process has led to profound changes in the arms industry of major European countries, and since 1989 it has become evident also in Italy.

The group pursuing this strategy is Finmeccanica, which has merged its two largest companies (Aeritalia and Selenia) and has acquired control or minority shares in many firms. The most recent acquisitions include 47 per cent of the shares of Elettronica, which Finmeccanica bought from Plessey when the British company was acquired by GEC and Siemens. As Ferranti was acquired in Britain, again by GEC, its Italian subsidiaries have also entered the Finmeccanica group. Minority shares in small private firms such as Magnaghi and Nardi were recently acquired by Finmeccanica. As Oerlikon (a Swiss firm in financial difficulty) is planning to sell its Italian subsidiary Contraves, Finmeccanica has already offered to buy it. An interest has also been expressed for Fiar, a subsidiary of the Swedish Ericsson group. It is remarkable that the only firms which could not be acquired by Finmeccanica are the aircraft and electronics companies of EFIM, the second state-owned holding, which is striving to defend its possessions.

While Italy's state industry seems to be entrenched in military production, a declining interest is evident in private groups, not only with foreign multinationals selling their Italian affiliates, but also with FIAT trimming its military production. In the early 1980s FIAT, with serious problems in its civilian production, rapidly increased its military activities by the acquisition of Snia and other firms, and by ambitious, but short-lived international initiatives, such as the joint venture with United Technologies in the take-over of British Westland. When the military market begun to shrink, the focus of FIAT efforts moved away from arms production. In recent years several reorganizations of the group's military activities have been attempted, all aiming at cutting jobs and losses and concentrating efforts in a few specialized niches.

The outcome of this process will leave more than one-third of Italy's arms production in the hands of a giant Finmeccanica. While the group's strategy is consistent with European patterns, its size would still be minor compared to the leading British, French and German firms. Furthermore, acquiring disparate companies in financial trouble does not mean developing an integrated business strategy, competitive at the European level. Such moves by Finmeccanica appear to come too late to reach a 'critical mass' in military production, and the framework of a credible, long-term industrial policy and co-ordination by the government is also missing, as the paths followed by the state-owned holdings clearly show. This costly wave of acquisitions at a time of declining markets, by a Finmeccanica which is already heavily indebted, is more likely to lead to a failed industrial strategy and to a government bail-out.

2. *The growth of international alliances.* Besides the growing national concentration, the European arms industry is marked by a rapid growth of cross-border acquisitions and international alliances among firms. Italian companies have remained largely outside this wave of cross-border agreements, shielded by a protected market and widespread state ownership.

One of the few steps in this direction is the joint venture started in 1991 between EFIM and the French group Matra Defense in missile production, but EFIM had first to create a new company, Oto Breda Missili e Sistemi, concentrating its activities in this field, previously scattered in three different firms.

The other forms of international agreement involving Italian firms remain far from the strategic alliances and the close integration which are now developing in several fields between divisions of the major European firms, such as British Aerospace, GEC, Aérospatiale, Matra, Thomson, Daimler Benz and Siemens. Italian firms have added to the number of firms involved in European co-production, linked to both NATO arms procurement programmes and to EC technology projects, but there is little evidence of wider-ranging inter-firm alliances. The acquisition of foreign firms has been limited to a few small US firms, in order to gain access to technology and new markets; examples include Aeritalia's share in the Dee Howard Co., a US engine firm, and Selenia's acquisition of Eyering Research.

The lag of Italian firms behind the development of European industrial alliances is an obvious consequence of the weakness of the Italian arms industry and can result in accelerating its decline. The firms' growing international orientation is the result of converging pressures from the demand and the supply side. The demand for arms procurement is increasingly affected by the role of the Independent European Programme Group (IEPG) in co-ordinating European arms programmes and by the slow steps towards more open national markets. With falling procurement budgets in all countries, the importance of international co-ordination is bound to increase, pressing for greater integration also at the industrial level.

On the supply side, the search for economies of scale and technological advances has favoured the concentration of Western Europe's arms industry in a few large firms with growing international alliances. The outcome of both processes is less scope for domestic production and national protection in countries such as Italy. The competitive and technological advantage of large international firms over smaller national companies will increase, and the policy of maintaining a subsidized and protected national industry through a growing arms procurement budget is likely to become unsustainable.

3. *Diversification in civilian markets*. Reducing a firm's dependency on arms production is a third common strategy in the current process of restructuring. Italian firms do not generally appear well prepared for a phase of declining arms spending and production; little planning for shifts to civilian products is visible, and most efforts have so far been devoted to lobbying for new arms procurement programmes. A few Italian firms, including Aeritalia, Selenia, Galileo, SMA and others, have explored the possibilities of a limited diversification towards civilian products as a way to preserve existing production capacity and fill the gaps between military orders. Most efforts have focused on disaster relief programmes, space and environmental monitoring projects or biomedical research. Sepa, Elsag and Ansaldo Sistemi have explored industrial automation systems, while SMA, a Florence electronics firm, has begun an interesting project to develop an anti-collision radar for motor vehicles.

Most of the diversification efforts by Italian firms have focused on short-term projects in technologies very close to military technologies; inevitably they have set their hopes for success on new, large government-funded programmes, which are the closest to military procurement.

The chances of a diversification strategy succeeding depend also on the specific industrial and market conditions in individual sectors. The aircraft industry has a great opportunity in the growing demand for civilian planes, and in the 1980s the Italian firms in this sector have joined several new international projects, from Aeritalia's participation in Airbus production to Aermacchi's agreement with Dornier to develop a new small civilian plane. In shipbuilding, the stagnating market for mercantile ships in Europe contributes to the prospect of a long-term decline of the industry. In mechanical and engineering production it is hard to identify growing markets where military producers could easily diversify. The recent slow-down in the automotive sector and the emerging difficulties for FIAT in Italy point out the problems such a strategy faces.

4. *Conversion to civilian production.* A more radical and long-term departure from arms production requires converting production capacities and human resources to civilian production. A growing interest in conversion is emerging in Italy, not only from trade unions and peace movements[9] but also from industry. In early 1989 the Minister for State Industry, Carlo Fracanzani, created a commission to study the conversion of military firms, whose report acknowledged the changes in the international scene and noted that there is 'no clear trend for the future on military sales, nor a well-defined firms' policy of conversion and diversification'. The report argued that state-owned groups and companies 'should propose in a short time a conversion plan to civilian production, based on their vocation, capacities and specialization', which could become part of the long-term programmes of the state industrial holdings.[10] However, as seen above, the strategy of the major state-owned military firms, and of Finmeccanica in particular, has gone in the opposite direction. Furthermore, in the summer of 1990 Fracanzani resigned over political differences with his party, and there was no follow-up to the work of the commission. A new conversion commission was soon established by the Ministry of Industry, including experts, representatives of firms and the military.[11]

In fact, the weakness of Italy's military firms is a major reason for directing the restructuring of Italy's arms industry towards a conversion strategy. In the present context of declining arms spending, a shift to civilian production of a large number of Italian arms firms, already extensively involved in civilian work, offers a more realistic way out of the current uncertainty, preventing the outbreak of a serious crisis as soon as past procurement programmes come to an end. Such a strategy would also be consistent with the established patterns of specialization in the Italian industry and, judging by past performances, it promises better results in terms of employment and growth. In this perspective, conversion represents an important opportunity for the restructuring of Italy's arms industry, reducing the size of a largely inefficient sector while avoiding losses of industrial capacity, and making new resources available for the civilian activities on which Italy's competitive success is based.[12]

[9] Among trade union initiatives on conversion, it is worth mentioning the inclusion of diversification clauses and limits on military production in company agreements bargained by the trade unions in firms such as Officine Galileo, Aermacchi, Agusta and others. Official statements, conferences and political campaigns have shown the interest of the Italian trade unions and peace movements in conversion issues. See Pianta and Castagnola (note 6).

[10] The report is quoted in Pianta and Castagnola (note 6), chapter 5.

[11] At the political level, the issue of conversion has been addressed by six proposals for new legislation, advanced by various political forces of the opposition. Most bills refer to the existing industrial and employment policy legislation and require the creation of a conversion fund at the Ministry of Industry, in order to finance the research, development, production and marketing of new products, while constraining firms to preserve their previous employment levels, and offering support for retraining programmes.

[12] In a conversion perspective, the European dimension again emerges as crucial, because of the large over-capacities in arms production in many European countries and the moves towards greater integration in production and procurement. Within the EC several important experiences of managed restructuring of traditional industries, from steel to shipyards, have been developed in the past, and the same tools of

The Italian arms industry is at a crossroads. Of the four different strategies outlined above, none has yet emerged as predominantly shaping the industry's future. While restructuring in Italy lags behind European changes, the delay in defining an appropriate and effective industrial policy for this sector is even greater.

Two opposite policies could lead the adjustment of Italy's arms industry in the 1990s. On the one hand there is the policy of upgrading arms production to the standards of major European countries, with a strategy requiring greater concentration and international alliances, and a continuing growth of arms spending and production, which seriously contradicts the present disarmament initiatives.

On the other hand there is the policy to shift the industrial capacity of arms firms to civilian production, through increasing diversification and eventual conversion. The strengths of such a strategy can be summarized as follows:

1. Unlike other West European arms industries, the Italian industry has so far suffered few major job losses and reductions, allowing a transition to civilian production to be planned before the emergence of an industry crisis.
2. The relatively small size and high diversification of Italy's arms firms would limit the economic impact of such adjustments.
3. Arms production in Italy is characterized by regional concentration in areas with strong industrial economies, experiencing rapid growth, where new skilled jobs can easily be offered.
4. The Italian technological specialization in areas relatively close to arms production, such as mechanical engineering and industrial automation, simplifies the transfer of the production resources of the arms firms into well-established and competitive civilian industries.
5. The importance of national arms procurement enables the Government to control the flow of resources to the arms sector, offering opportunities to shift public spending to other areas, as in a policy of demand-oriented development of new civilian products.
6. The large share of state ownership means that the Government could, with an appropriate industrial policy and company strategy, dismantle a significant part of the Italian arms industry and expand civilian work.

For all these reasons, the restructuring of the arms industry in Italy could have a different outcome from the industry-led concentration and internationalization of arms production evident in other major European countries. The size and characteristics of Italian firms facilitate a strategy of diversification and conversion, which would be not only socially desirable, but also less painful for workers, communities and firms. It remains to be seen whether this opportunity to apply such an appropriate and timely policy will be taken to direct the current restructuring of the Italian arms industry.

industrial policy and co-ordination could be used to manage the reduction of the arms industry. Within the European Parliament and trade unions, an interest in a strategy of this kind has already emerged.

13. Arms production and modernization in Spain

Jordi Molas-Gallart[*]

I. Introduction

In the 1980s the Spanish defence industrial policy underwent a complete revision. The Socialist Government, elected in October 1982, perceived a need to modernize the obsolete Spanish military industrial base and to raise it to a 'European' level. It saw modernization as the way to overcome the questionable management procedures and the endemic losses that characterized the Spanish military industry. Investment in research and development (R&D) and participation in international arms development and production programmes have been the main tools used to upgrade the technological capabilities of the Spanish defence industry.

The growth strategy for the Spanish arms industry relied heavily on the access to foreign (Third World) markets. After 1986, when Spanish arms exports declined, the arms industry found itself in a difficult position. New markets had to be opened to compensate for the deterioration of exports to the Third World. To find new markets in developed countries, the technological level of the products offered had to be improved. This improvement in quality required high fixed investments that had to be compensated with increases in the production runs. Given the limits of the domestic market, an increase in production meant an increase in exports; more exports were therefore needed to compensate for the collapse of the export market.

Spain's answer to this situation lay in intensifying the policy of participation in international arms projects and increasing the defence research effort. It is expected that international projects will allow Spain access to technologies that would otherwise be beyond its reach and offer new markets to the Spanish arms-related industry. To help justify the growing investment in an area of dubious market prospects and strategic rationale it is argued that military R&D will be beneficial for the development of key high-technology sectors in the Spanish industry.

The possibilities for a diffusion of technologies and incentives for technological innovation from the military-related to the civil areas of economic activity depend largely on the structure of the Spanish military-related industry. This chapter undertakes a descriptive analysis of this structure. In order to

* Financial support for writing this paper was provided by a NATO fellowship.

analyse the links between military-related economic activities and the rest of the economy it is very important to distinguish between the different levels in the production chain from components to final integrated systems. This chapter advances a new technique of analysis based on a definition of military production and a corresponding classification of military-related (MILREL) firms. This taxonomy is interfaced with sectoral and geographic classifications which together offer a map of the structure of the Spanish military-related industry.

The definition and taxonomy adopted are addressed in sections II and III. Section IV provides an analysis of the geographic distribution of military-related plants in Spain, and section V addresses the structure of suppliers and buyers in the military-related industry.

II. Definitions

Conceptual problems have hindered an assessment of the economic importance of the military industry. The armed forces consume an enormous variety of goods sometimes similar or equal to those used in the civil area. The specific needs of the military are not constrained to arms and weapon systems. Too often the study of the economic role of military-related production has been limited to the study of arms production. There are two main difficulties with this approach. First, the arcane nature of the concepts 'arms' and 'weapon' has proved to be a source of confusion.[1] The distinction between 'arms' and products and technologies for (or relevant to) military use is blurred. Second, to limit the analysis to 'arms' production obscures the vertical integration (and hierarchy) of military-related production. This distinction is important for an understanding of the relationship between military-related industry and other economic activities. By including the production of subsystems and components in the analysis, it undertakes to consider an area of industrial activity where the differences between military and civil products may be less apparent and the potential for cross-fertilization is higher. Therefore, any analysis of military production that leaves out the producers of subsystems and components cannot present a proper picture of the role of military production in technological innovation.

To allow the inclusion of subsystems and components, a military product is defined as any item with specific characteristics primarily geared to satisfy the demand of the armed forces and embodying, therefore, a certain degree of military specificity. This analysis considers all firms engaged in the manufacture or design of military products, that is, all military-related companies. This encompassing definition of 'military product' enables the analysis to include all military-related industrial activity.

[1] This difficulty has been illustrated by the legal rows arising from legislation banning or controlling 'arms' exports.

III. Classification

The broad nature of 'military production' creates a new problem for the researcher. Firm size, dependence on defence demand and complexity of the products manufactured differ widely among military-related companies. The problem is therefore how to account for this diversity when much of the available information is supplied by the firms themselves.

One way is to classify firms according to the products they manufacture. Four levels of military product are distinguished and ordered according to their position within the production ladder,[2] and final products are ranked based on their use in what is a hierarchically integrated system: the armed forces. Thus, weapon systems are ranked above subsystems and components. Auxiliary *matériel* such as uniforms or mobile kitchens are final products which are not as central to military activity as arms and munitions; these are therefore ranked in a 'lower' group. Major weapon systems, being the axis around which most of the military tactics and strategy revolve, are accorded the highest level in this hierarchical classification. The main products included in each level of Spanish firms can be grouped as follows:

A. Major integrated weapon systems. These include all types of military aircraft, as well as tanks, aircraft-carriers, frigates and submarines.

B. Arms and munitions, missiles, special and armoured vehicles; electronic warfare, integrated radar and early-warning systems; advanced major weapon system simulators; and design and modernization of major weapon systems.

C. Land transport vehicles, fast boats; subsystems, communication kits; fire control, range finders and night vision subsystems; engines; weapon and weapon system maintenance; design of level B items; and auxiliary products of military character (helmets, uniforms, tents, mobile kitchens, flares, signal equipment, etc.).

D. Components, process machinery (when of special character) and other special materials.

Included in level B is the design of level A products. Sometimes the design of a product or the software needed to operate it is done outside the main manufacturing firm. These stages are increasingly important in the development and manufacture of military products and should be included in any assessment of the importance of military production for technological innovation. As they are

[2] In the Spanish case, the 'top' of the domestic production ladder does not correspond to the top of the armed forces equipment structure. The Spanish aeronautics industry manufactures a jet trainer and transport aircraft but not advanced fighters. Since the major objective of this research is to examine the relationship between military-related firms and the rest of the economy, the dominant discriminating factor is the position of the product within the domestic industrial hierarchy. The same criterion will lead to different classifications in countries with different production structures and is, therefore, not useful for comparative analysis.

from a previous stage in the production of a final product they are included in the level immediately below that of the designed or 'programmed' item itself.[3]

This categorization by product enables a classification of firms. Each firm's level is that of the highest level military item it has proved to be able to manufacture.[4] For instance, if a firm manufacturing mainly electronic components (level D) also produces a fire control system (level C), it will be considered as belonging to the level C group, regardless of the relative significance of this production in the overall activity of the firm. The underlying rationale for this criterion lies in the technological ability and capacity needed to produce goods of a certain level in this hierarchy, regardless of the relative significance of such items in a firm's total range of products.

The firms producing items in level A have to acquire key skills in the area of systems integration and are likely to become prime contractors in the production of integrated arms systems.[5] The machinery, skills and marketing strategies needed to produce, assemble and sell major weapon systems differ from those required by firms operating at lower levels of this hierarchy. The higher a military product is in the hierarchy, the more complex it is, and the more dissimilar it is likely to be to products of civil industry. Conversely, the lower the level of the military product, the more similar it is likely to be to products of civil industry, or the easier its application to civil uses.

Firms are also classified according to industry sector (electronics; software and systems engineering; metal; vehicles; aerospace; chemical; textile and other; and shipbuilding). Spanish firms rarely operate in more than one sector. In the few cases that this happens, the company is classified within the sector in which its production is concentrated.

IV. Geographic and sectoral distribution

This section presents an analysis of the geographic and sectoral distribution of military-related plants in Spain. Public sources were used to compile the list of 419 firms used in the study. The section also presents an analysis of the structure of suppliers and buyers, on the basis of confidential data provided by 200 firms to the Ministry of Defence during 1986 and 1987. The availability of data constrains the choice of base years.

It is assumed that an even geographic distribution will improve the chances for the diffusion of technological incentives beyond the military production

[3] For example, the design for an aircraft-carrier (level A) will be considered as a level B product, and the software to control a heavy gun (level B) will be considered as a level C product.

[4] A firm is considered to manufacture a product when this product is on offer in the reference year (1986) or has been produced in previous years and could be produced again if the demand existed. Trading firms and management consortia are excluded from this analysis.

[5] The central role of the 'system integrator' may be receding, giving way to system designers and some key component producers as prime contractors. This is a very recent development which has so far not had a noticeable effect on the Spanish military industry, although there are some indications of changes taking place. For example, SENER, a systems engineering firm, has just moved into manufacturing by becoming the most important participant in a new company set up to manufacture parts for the European Fighter Aircraft (EFA) jet engine (see note 10).

Table 13.1. Plants by province and production level (A–D), 1986–87

Province	A	B	C	D	Total
Madrid	2	13	60	97	172
Catalonia	–	–	24	70	94
Basque Country	–	20	15	50	85
Rest of Spain	7	26	46	64	143
Total	**9**	**59**	**145**	**281**	**494**

Source: Author's calculations from Spanish Ministry of Defence data.

area. If military demand is a potential source of technological incentives, the diffusion of these incentives to the rest of the economy will be favoured by the dispersion of production plants. Alternatively, the concentration of military-related plants may suggest a production structure revolving around itself and secluded from other economic activities. The extreme case would be an 'enclave economy', where a military industrial sector exists in isolation from the rest of the economy, and where industrial development in the military industry does not trickle down to other sectors. High-technology military industries may potentially develop into 'enclave economies' if the domestic industrial structure cannot provide an industrial base with which the military firms could establish backward linkages. An analysis of the distribution of plants may show the non-existence of an enclave and provide additional insights into the structure of the Spanish military-related industry and its potential as a technological spearhead for the rest of the economy.

Four areas of distribution are distinguished in this study: the three most important industrial regions (Catalonia, the Basque Country and the province of Madrid) and the rest of Spain. The latter includes mainly agricultural regions and localized industrial zones such as Zaragoza, Valladolid and Seville.

Interestingly, the production of arms and weapon systems is particularly important in the rest of Spain, that is, outside the leading industrial regions (see table 13.1). Conversely, in the traditional industrial areas there is an important presence of firms at level D: 57 per cent of the listed plants in Madrid, 74 per cent in Catalonia, and 59 per cent in the Basque Country are operating at level D. In the rest of Spain the percentage of components producers is only 45 per cent. The rest of Spain accounts for an important share of the production of arms (levels A and B) but does not fare particularly well in the production of subsystems and components. This may suggest the failure in the creation of a local components supply base for the arms industry when the latter is located outside traditional industrial areas. In contrast it highlights the capacity of the traditional industrial zones to satisfy at least some needs of the arms industry. Therefore, in aggregate terms, the non-existence of an enclave economy can be asserted, as there is a strong conglomerate of subsystems and component firms in the traditional industrial regions producing for the military industry.

Level A and many level B companies are publicly owned. These include the Bazán shipyards, the aerospace firm CASA and all the arms factories previously belonging to the armed forces and now integrated within Santa Bárbara.[6] Their location is largely the result of historical events and decisions taken by the state. Many plants were established during periods of armed conflict, when the choice of location was determined more by the distance to battle fronts and easily targetable industrial concentrations than by infrastructural needs. Furthermore, industrial concentrations have in Spain historically been centres of social unrest that could have posed a threat to such a 'sensitive' industry. Consequently, arms factories were spread over the national territory, often far from the main industrial regions.

More detailed sectoral analysis shows the diversity in the structure of military industry. The most important differences appear between traditional sectors such as the chemical, metal, textile and shipbuilding industries and the newer 'high-tech' industries in areas such as aerospace, electronics, and software and systems engineering.

The distribution of plants in the metal sector highlights the position of the Basque Country as the most important region in military-related production in this sector. In fact 80 per cent of Basque plants at level B and more than 60 per cent at level D are in this sector. The high number of Basque plants at level B denotes the prominence of the Basque arms manufacturing industry, composed of medium- and small-sized companies.[7] However, this is not translated into a presence in the sphere of heavy arms manufacturing;[8] even at the level of components and sub-assemblies, the presence of Basque firms is small.

The presence of arms producers in the metal sector is also very important in the rest of Spain. The only plant that has manufactured tanks is in Seville. Yet, despite the important number of arms manufacturing plants in the rest of Spain, relatively few factories producing components are located in these regions. For instance, Catalonia has almost the same number of MILREL component manufacturing plants in the metal sector, despite the fact that it has no level A or B plants.

The sectors for electronics and for software and systems engineering are central to defence industrial policy as well as to global science and technology policy. The Government's R&D and industrial policies attach special importance to the electronics sector and have used the promotion of defence electronics as a central policy tool. The analysis reveals a heavy concentration of plants in this sector in the Madrid area. More than three-fourths of all

[6] Bazán is the abbreviated name of the dominant, state-owned, military shipbuilding firm Empresa Nacional Bazán de Construcciones Navales Militares, S. A.; CASA is the acronym of the public Spanish aerospace firm Construcciones Aeronáuticas, S. A.; and Santa Bárbara is the short name for Empresa Nacional Santa Bárbara de Industrias Militares, S. A., a public firm set up in 1959 to absorb all the manufacturing facilities then under the direct control of the Army. The process of transfer was slow and lasted from 1960 to 1987.

[7] This is also reflected by the presence of many level B chemical plants (munition factories) in the Basque Country.

[8] It is also interesting to note the absence of MILREL shipbuilding plants in the Basque Country despite the importance of this sector in the Basque economy. Basque shipbuilding is almost exclusively civil.

software and systems engineering firms and almost two-thirds of the electronics plants are in the province of Madrid.

This concentration is even more evident in those electronics firms placed higher in the hierarchy. All electronics plants producing level B items are located in the province of Madrid. The importance of these electronics firms is increasing, in terms of both military sales and the role played in the Spanish military industry as a whole.[9]

The distribution of the plants involved in the manufacture of electronics components conforms to the general pattern of a relative absence of components manufacture in the rest of Spain and a specialization of Catalan plants in the production of components. Most Catalan electronics firms produce only components (level D) whereas in the rest of Spain the number of plants producing at level C is double those at level D, suggesting a lack of industrial support in components manufacture.

The aerospace sector is another high-technology area receiving special Government attention. Very few firms specialize in aerospace production. Although some suppliers are found in other sectors (e.g., avionics suppliers are listed under electronics), the layer of domestic suppliers for the Spanish aerospace industry remains thin. Managers of the main Spanish aerospace company, CASA, have long complained about the lack of an industrial base of subcontractors capable of supplying the needs of the aerospace sector.[10]

The only Spanish aerospace factories are in the Madrid area and in the southern region of Andalusia, where CASA has three plants.[11] The concentration of level A plants in the Seville–Cádiz region is a noticeable characteristic of the Spanish military industry. Apart from the CASA plants, Seville also hosts the only plant at level A in the metal sector,[12] while Cádiz hosts a Bazán shipyard (also level A). Predictably, this concentration in the area of heavy armaments production does not translate into a similar strength at the lower levels of the production hierarchy.

V. Suppliers and buyers

Table 13.2 gives an indication of the structure of the suppliers to and buyers from Spanish military-related firms, by industry sector and production level. The data for this table were provided by the Spanish Ministry of Defence, based on information drawn from replies to questionnaires received from 200 firms.

[9] In contrast, the Basque Country is clearly specialized in the metal sector, with a conspicuous absence of MILREL electronics plants. Therefore the sectoral shift in military production implies a change in the geographic balance of military production.

[10] CASA has since made public its intention to help to create a network of subcontractors to provide a domestic supply for certain high-technology components. In 1989 CASA reached an agreement with the British company Lucas Aerospace to create a company to produce hydraulic components for the aerospace industry, and has participated in a new aerospace engine company (Industria de Turbo Propulsores) together with the Spanish firm SENER and British Rolls Royce to produce some parts of the EFA engine.

[11] Industria de Turbo Propulsores (see note 10) will be located in the Basque Country.

[12] This is the Seville Artillery Factory, one of the most important Santa Bárbara plants and the only one where tanks have been assembled.

Table 13.2. Suppliers to and buyers from Spanish military-related firms, by industry sector and production level (A–D), 1986–87

	Elec.	SSE	Met.	Aer.	Veh.	Tex.	Ship.	Chem.	A	B	C	D
Supplier												
Foreign[a]	53	33	40	9	8	6	7	7	5	30	65	79
Domestic[b]	58	20	232	1	35	18	17	49	4	71	115	292
Buyer												
Foreign[a]	11	9	17	3	–	1	–	3	3	16	8	21
Domestic[b]												
Public[c]	30	8	27	4	7	3	2	6	3	25	29	34
Private[d]	62	44	163	–	9	8	8	32	–	38	99	207
Level												
A–B	29	29	73	5	6	3	4	25	4	57	46	63
C–D	16	6	32	–	2	2	2	2	–	5	13	42
E	46	23	92	1	9	6	4	17	1	9	72	143

[a] Foreign factories and institutions (foreign-owned factories in Spain are not included).
[b] Spanish factories or institutions.
[c] Spanish state agencies, including those of the Ministry of Defence and the armed forces.
[d] Spanish private sector.
Elec. = Electronics; SSE = Software and systems engineering; Met. = Metal; Aer. = Aerospace; Veh. = Vehicles; Tex. = Textile and other; Ship. = Shipbuilding; Chem. = Chemical. A, B, C, D, E : production and buyer level groups as described in the text.

Source: Author's calculations from Spanish Ministry of Defence data.

The firms listed their four most important buyers and suppliers. To protect confidentiality, aggregated data are used. The sectoral and level classifications given above are used with a slight modification. Buyers that are military-related firms are classified as above (levels A, B, C and D). Firms that are not military producers are included in a new level: E. Final consumers (armed forces, ministries of defence, ministries of the interior and other public clients) are classified as follows:

Levels A–B: armed forces, ministries of defence and other ministries when the object of transaction is arms for the security bodies depending on them.

Level E: all institutions that are not related to military, paramilitary or security activities.

In the first place the data in table 13.2 point to the consistency of the classification by level groups. Of the most important clients of firms at the A and B levels (the core of military production), 80 per cent are either other arms manufacturers (same level group) or armed forces and defence ministries. This percentage diminishes down the production hierarchy. For firms at level C (subsystems and auxiliary *matériel*), 35.1 per cent of their most important clients are producers or consumer of arms and military systems, and for the firms at level D (manufacturers of components) this percentage is only 25 per

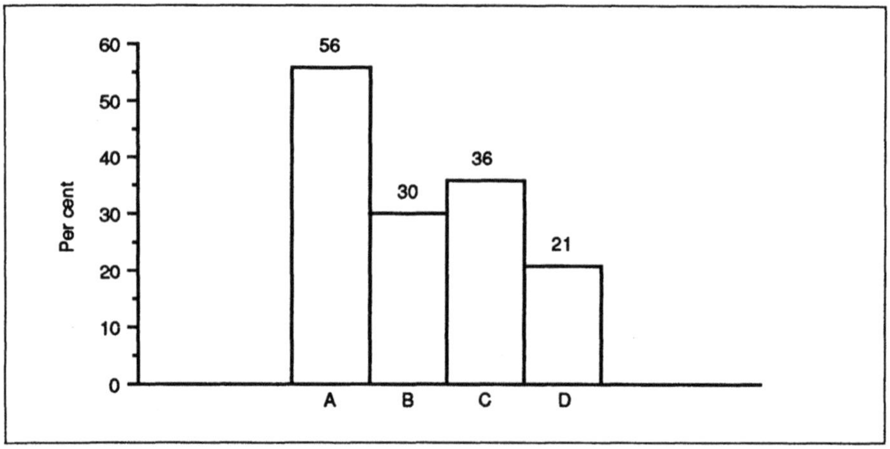

Figure 13.1. Percentage share of foreign suppliers to Spanish military-related industry, by production level (A–D), 1986–87

Source: Author's calculations from Spanish Ministry of Defence data.

cent. This is still an important share that reveals the relationship between components manufacturers and military production, but it is very low if compared with producers at levels A and B. Component manufacturers enjoy a more diversified trade, with most production earmarked for civil markets but with a sizeable military-related market. This suggests, as one would expect, that firms lower down in the production hierarchy have a larger percentage of buyers outside military-related production.

Suppliers

In terms of suppliers, important differences appear depending on the sectoral affiliation of the MILREL firms concerned. There is an acute reliance on foreign suppliers in the newer high-technology sectors (electronics, aerospace, and software and systems engineering), while the number of foreign suppliers in other sectors is comparatively low. For some traditional sectors, such as metal, the percentage of foreign companies among the most important suppliers to the firms is as low as 15 per cent (see figure 13.1).

The case of the aerospace sector is the most remarkable, with only one domestic supplier among the 10 cited.[13] This adds further empirical evidence to the argument that the domestic industrial base is not sophisticated enough to support the aerospace sector. The lack of domestic suppliers to the Spanish aerospace sector also indicates a shortage of channels for the 'downward' diffusion of technological incentives. Furthermore, only in the aerospace sector has it always been possible to identify main activity on the supplier side. The

[13] The military-related aerospace sector includes the main firm in the sector, CASA, and a small aerospace auxiliary firm.

Table 13.3. Foreign and domestic firms among the main suppliers for firms in the electronics sector, by production level (B–D), 1986–87

Level	Foreign	Domestic
B	15	2
C	20	22
D	18	40

Source: Author's calculations from Spanish Ministry of Defence data.

Spanish aerospace industry seems to have a predominant role as a subcontractor to foreign aerospace companies.

In the case of electronics, the concentration in Madrid of MILREL firms and the high number of foreign suppliers point to a weak relationship between this sector and the rest of Spanish industry. Moreover, as seen in table 13.3, the proportion of domestic and foreign suppliers to electronics companies changes dramatically according to the firms' level. Domestic suppliers become more important in the lower levels of the production hierarchy. The significance of the difference between levels cannot be underestimated. The chance for diffusion of technological incentives is much greater at level D than at level B. The weight of foreign suppliers for electronics firms operating at level B combines with the concentration of such firms in the province of Madrid.

Given the reliance of Spanish high-technology MILREL firms on foreign suppliers and technologies, there is a risk that the Government's incentives to stimulate the growth of high-technology industries, through military procurement orders, will simply be exported. Congruent with the preponderance of foreign suppliers in the Spanish high-technology industry is the widely held belief that Spanish MILREL electronics and aerospace firms are limited to assembling or manufacturing of low- to medium-technology components and subsystems. Foreign firms also often appear both as main suppliers and as main buyers. In such instances the role of the Spanish firm is reduced to that of a subcontractor assembling components for a foreign company.

That the presence of national suppliers is the highest in the metal, vehicles, chemical and 'textile and other' sectors comes as no surprise. These are traditional sectors with a predominantly low- to medium-technological level[14] and a long-established and important presence in Spain. Therefore, the sectoral traditions and strength of the Spanish industry seem to be reflected in the structure of suppliers to the military industry.

Reliance on foreign suppliers varies according to the level of manufactured military products not only in the electronics sector but also when all military-related firms are considered in an aggregated form. Figure 13.1 shows a clear difference between level A and the other levels as regards the importance of foreign suppliers. Foreign suppliers are more important to firms producing

[14] The chemical sector may be an exception in this respect.

weapon systems and less important to firms producing only components. The producers of subsystems (level C) and of arms and munitions (level B) are in an intermediate position.[15]

The stronger presence of foreign suppliers among producers of weapon systems reflects the greater complexity and specialization of production and integration on this level, requiring Spanish system integrators to turn to foreign suppliers which can fill this need. By contrast, firms whose military production is limited to component manufacture are more closely integrated in the national industrial structure, through a larger number of domestic suppliers. This may be attributed to a lower level of complexity on this level of production, both in terms of the finished products and in terms of *matériel* requirements.

Buyers

The analysis of the most important clients of the military-related companies is also revealing (see table 13.2). An important aspect of the buyers' structure of MILREL firms is the prominence of public clients in the aerospace and, to a lesser degree, in the vehicles and electronics sectors. About one-third of the main buyers in the MILREL electronics sector and one-half in the MILREL aerospace and vehicles sectors are also in the domestic public sector.[16] Clearly, the vehicles and the electronics sectors are not comparable. Although the most important clients for firms in the vehicle industry may be in the public sector, production is mainly targeted to small private buyers. Vehicle producers are usually large, with sufficient capacity both to satisfy defence needs, thereby becoming important military suppliers, and, at the same time, to concentrate efforts on the private civil markets. Electronics firms tend to be much smaller and more specialized, wherefore the relative importance of sales to public agencies, including the military, tends to be much greater. This is particularly the case in Spain, which lacks a large private market for domestic electronics.[17]

The considerable share of public agencies among MILREL aerospace and electronics buyers reflects a high degree of specialization, indicating that the capacity of these sectors to adapt to the competitive environment of private markets may be impaired. In the Spanish case, the high-technology MILREL industries may become dependent on orders from the public sector, and thus become increasingly secluded from international competition and from expanding private markets. This would limit their ability to achieve product diversification and a transfer of products and processes from public (including military) to private (civil) markets.

[15] The sectoral composition of the groups affects this distribution. The presence of foreign suppliers is slightly higher, in relative terms, at level C than at level B due to an important presence of electronics firms at level C and of metal and chemical companies at level B.

[16] Foreign states are excluded.

[17] Spanish MILREL electronics firms seldom produce consumer electronics.

VI. Conclusions

This chapter suggests a technique of analysis and highlights some of the characteristics of Spanish military-related industry. Further research is needed to complete the picture. Some tentative conclusions may nevertheless be drawn.

1. The different indicators point to some perturbing traits of the aerospace, electronics, and software and systems engineering sectors. As seen, the firms in these sectors rely to a great extent on foreign suppliers and public clients, and they are geographically clustered in and around Madrid. This indicates a certain degree of isolation of the military-related high-technology firms from civil markets and from the rest of the economy.

2. The overall degree of domestic industrial support to military production in the metal and chemical sectors is higher than in the newer high-technology areas, reflecting a continued Spanish manufacturing tradition in the former. At the same time, Spanish defence industrial policy has recently been focused on aerospace and information technology, which is deemed to be central to the present technological revolution. When R&D contracts and other incentives to technological change are offered to firms in other sectors, they are targeted to the companies at level A, that is the major state-controlled arms producers.[18] However, these companies have a greater tendency to resort to foreign suppliers than the firms operating at other levels. Foreign suppliers are most important to firms producing complex weapon systems (level A) and least important to firms producing components (level D). The producers of components and subsystems are also more dispersed geographically, their number is larger and their dependence on sales to the military is lower.

The Ministry of Defence has focused its science and technology policy on the high-technology sectors and on producers at the higher levels of the hierarchy, on the assumption that they will constitute a spearhead of technological development in the quest to upgrade the technological level of Spanish industry. However, these hopes may be misplaced if the apparent tendency of the high-technology military-related sector to remain isolated from the rest of industry is reaffirmed. There is the possibility that the technological incentives and technological change thus initiated will remain within a reduced group of MILREL firms. Planners of industrial defence policy should be aware of this possible shortcoming and closely monitor the relationship between MILREL firms and their economic environment. The incentives for technological development provided by the military authorities may have greater effect if targeted at those companies more deeply rooted in the Spanish industrial base.

[18] Firms at level A (including CASA) and medium-size electronics firms share most of the Ministry of Defence R&D funds.

14. The developing arms industries in Greece, Portugal and Turkey

*Anthony Bartzokas**

I. Introduction

In scenarios for further integration of arms production in Western Europe, small European NATO countries, especially those in southern Europe, play a limited role. Their marginal technological capabilities and lack of political influence, in conjunction with their small internal arms markets, do not provide serious opportunities for significant involvement in the bargaining process. Since the 1950s the so-called DDIs (NATO countries with developing defence industries, i.e., Greece, Portugal and Turkey) have participated in the European armament process in four different ways. They have: (*a*) obtained used military equipment through military grants; (*b*) bought arms on the international market; (*c*) bought military manufacturing technology; and (*d*) sought to play an active role in current European attempts to create an integrated arms market. The purpose of this chapter is twofold: to present briefly the patterns of military industrialization that the DDIs have pursued, and to examine the current prospects of their respective armament sectors. Two aspects of the international context will be introduced. The first is historical: the interrelation between major European arms manufacturers and the local defence market. The second is actually a set of factors: the reconstruction of the European military industry, the urgent need for new markets and the current surplus of arms in Central Europe.

II. Defence market interrelations

For a long period the DDIs have been net importers of military equipment from allied countries, mainly the Federal Republic of Germany and the USA. During the 1950s equipment from World War II and the Korean War constituted the main part of these transfers. This military aid programme was part of a broader plan for the integration of the smaller NATO countries in the allied structure of armed forces and production capabilities. According to NATO plans, countries such as Greece, Turkey and Portugal should retain large armed forces equipped with low-cost equipment provided by their major allies. The expectations for

* I would like to thank Constantine Vaitsos and William Walker for valuable comments. Financial support was provided by the Greek Institute for International and Security Studies and the British Social and Economic Research Council. The views expressed are those of the author.

Table 14.1. Value of major NATO countries' military exports to Greece, Portugal and Turkey, 1963–87

Figures in roman are in US$ m., current prices; figures in italics are percentage shares of European and world sales, respectively.

Country	1963–73	1973–77	1978–82	1983–87
FRG	295	320	850	1 275
European sales	*38*	*54*	*61*	*47*
World sales	*25*	*14*	*15*	*16*
France	183	365	190	60
European sales	*26*	*31*	*21*	*15*
World sales	*7*	*8*	*1*	. .
UK	. .	. .	25	470
European sales	. .	. .	*3*	*34*
World sales	. .	. .	. .	*1*
Italy	. .	290	130	70
European sales	. .	*80*	*48*	*28*
World sales	. .	*25*	*3*	. .
USA	2 194	2 364	2 550	2 315
European sales	*22*	*43*	*22*	*14*
World sales	*7*	*9*	*7*	*4*

Source: US Arms Control and Disarmament Agency, *World Military Expenditures and Arms Transfers* (US Government Printing Office: Washington, DC, 1989), table III.

the development of indigenous military industries were therefore limited. Instead these countries gave priority to the creation of maintenance networks for the imported weapon platforms. They also developed limited capabilities in ammunition, aircraft maintenance and, in the cases of Portugal and Turkey, naval shipbuilding and aircraft assembly.[1]

This picture changed in the mid-1960s with the elimination of ageing military equipment from US and other NATO inventories. Gradually, military grants decreased and the DDIs became a significant export market for arms manufacturers in Europe and the USA. Table 14.1 highlights this trend for the major European countries. The FRG is well established in the DDI market (accounting for almost half of its European military exports), while the UK expanded rapidly in the 1980s, almost exclusively with exports to Turkey.

Of particular relevance is also the deterioration of Greek–Turkish relations since 1974. In the ensuing regional arms race both countries rapidly increased

[1] For Greece, see Bartzokas, A., 'Military technology transfer and domestic arms production: the case of Greece', thesis presented towards doctoral degree, University of Sussex, 1991, chapter 3; for Portugal, see Ulling, B., 'Portuguese Air Force', *Aircraft Illustrated*, vol. 20, no. 4 (1990), pp. 190–91; and BDMI, *The European Defence Dossier*, vol. 3 (BDMI: Southampton, 1990), p. 1024; for Turkey see Akgul, A., 'Turkey's defence industry: ambitious for growth', *Asian Defence Journal*, vol. 18, no. 10 (1988), p. 106.

their defence budgets.[2] Since 1975 these two countries have been a significant market for advanced weapon systems. Moreover their policy of reducing dependence on monopolistic suppliers gave to West European arms manufacturers the opportunity to gain access in those markets and eventually to increase their shares. By contrast, the Portuguese defence procurement process faced a long-standing stagnation due to the end of its colonial commitment and to the transfer of resources to other areas of central government expenditure.[3]

The implementation of large-scale procurement programmes created a serious handicap to the economies of these countries.[4] However, the foreign trade deficit, shortages of foreign currency and devaluation of the national currencies against the dollar have been constant structural problems. Under these circumstances, the promotion of military exports in the DDI market was related to credit supplies from the US Foreign Military Sales (FMS) programme and, on a case-by-case basis, from France, the FRG and the UK. Also the assimilation of common military doctrines and operational requirements, through the involvement in the allied military hierarchy, stimulated preference for specific types of weapon systems. Generally speaking, the establishment of constant channels of equipment supply and the operation of mechanisms for the transfer of military know-how had crucial importance at the micro-level of the decision-making process.[5] Of course, the contemporary NATO structure does not include an integrated defence planning and management process. However, the adoption of common allied perceptions could be decisive, especially in instances when the procurement authorities have to change the general operational plans to accommodate specific military articles.[6] The historical experience of intergovernmental collaboration between the DDIs and advanced industrialized countries demonstrates that these relations are closely related to state/corporate efforts to export weapons to the DDIs. State-to-state memorandums of understanding signed by DDI governments and major allied nations have regulated the transfer of second-hand equipment and other compensation

[2] For a description of the foundations of the local arms race, see Giannitsis, A., *Ellas–Tourkia: Oikonomikes Apopseis ton Exoplismon* ['Greece–Turkey: economic aspects of armament expenditures'], *Epitheorisis Koinonikon Erevnon*, no. 25 (1975), pp. 425–38 (in Greek). For an interesting theoretical perspective, see Deger, S., 'Models of arms race among neighbouring countries in Less Developed Countries', *Discussion Papers in Economics*, no. 119 (Birkbeck College: London, 1982).

[3] See Daguzan, J. F., 'Les politiques industrielles de defence de l' Europe du Sud', *Defence Nationale*, vol. 44 (Dec. 1988), pp. 110–11; Hewish, M., 'Portugal's equipment priorities', *International Defence Review*, vol. 20, no. 12 (1987), p. 1663.

[4] According to econometric research by Antonakis and Karavidas, over the period 1950–85 military expenditures cost Greece a fall in the annual growth rate of the economy by 0.52–0.62 per cent per year. See Antonakis, N. and Karavidas, D., 'Defence expenditures and growth in LDCs: the case of Greece, 1950–1985', Paper presented at the 3rd Congress of the European Economic Association, Venice, Aug. 1988, p. 21.

[5] For a discussion of the micro-level, see Danopoulos, C., 'Alliance participation and foreign policy influence: the military's role', *Armed Forces and Society*, vol. 17, no. 2 (winter 1985), pp. 271–90.

[6] For a discussion of the technical assistance programmes to the DDIs as a way of cost-effective defence aid, see Rupp, R., 'Assistance to Greece, Portugal and Turkey: a form of burden-sharing', *NATO's Sixteen Nations*, vol. 30, no. 7 (Dec. 1985), pp. 45–58; Rupp, R., 'Defence-industrial assistance to Greece, Portugal and Turkey', *NATO's Sixteen Nations*, vol. 32, no. 6 (Oct. 1987), pp. 81–88.

agreements. Terms of loans are a significant input in the decision-making process for new orders. The integration of these countries in the allied structure had a market-oriented form. Defence needs were covered through imports from advanced allied countries. Permanent links were established between military markets and major Western manufacturers. The main body of defence aid was concentrated on the implementation of import contracts, not on the creation of indigenous manufacturing capabilities.

In the 1980s the prospect of an integrated European military market introduced a new element to the traditional role of DDIs: the possibility to take part in an integrated European arms sector. Local firms could benefit from joint programmes and increase their exports to other West European countries. Every proposal for a more integrated West European arms industry opens up possibilities for political pressure for market preferences. The preferred model for the DDIs is some kind of flexible complementary participation in the integrated West European armament process. A major problem is the well-established dependence on military imports from the USA. The USA has used military aid funds, knowledge transfer, cheap prices and direct political pressure in order to secure sales. Under these circumstances, the DDI countries could decide to move towards a fully integrated European military market only with difficulty.

In 1987 as a response to a report prepared by the Greek delegation, the Independent European Programme Group (IEPG) agreed to increase defence aid programmes. In the future the DDIs' defence equipment spending, whether in collaborative projects or in bilateral contracts, should also be more or less balanced against the return they receive in accordance with the principle of *juste retour* (see also chapter 3). This is in fact the core of the IEPG's proposal for the DDIs. However, the implementation of this proposal raises serious and complicated problems of initial definition, such as regarding types of formal obligation, categories of equipment for buy-back orders and contractual prices. Meanwhile European governments and leading industrialists are sceptical about this formula, which they consider static and inefficient.[7] At any rate, the prospects of compensation returns to the DDI depends mainly on a political decision, and it remains to be seen what will be the final compromise between problems arising from expanding military manufacturing capacity and access to the DDI market. In the late 1980s, the DDIs' procurement budgets have experienced slow rates of growth or even a decline (see appendix C). Leaving aside the problems arising from the devaluation of national currencies, they

[7] According to Mr Dassault, President of the French aerospace company Dassault: 'Altogether it seems very difficult to reconcile the maintenance of the principle of fair return in collaborative procurement programmes with the search for rationalization, more efficiency and the greatest possible savings'. Cited in Western European Union, *European Co-operation in Armaments Research and Development: Guidelines Drawn from the Colloquy*, 34th Ordinary Session, Doc. no. 1141 (WEU: Paris, 10 May, 1988), p. 11. Until now, the involvement of DDIs in co-production programmes has been limited to the Stinger project (involving Greece and Turkey). See 'Dornier leader of the Stinger licence production in Europe', *NATO's Sixteen Nations*, vol. 34, no. 3 (June 1989), pp. 84–88. The involvement of DDIs in the EUCLID programme has already raised complaints of 'free riding'. See Reed, C., 'EUCLID: the future of European defence cooperation', *Defence*, vol. 21, no. 6 (June 1990), pp. 344–48.

have to solve serious economic problems in their attempt to make the required macro-economic adjustments. Even Turkey, which has had the most ambitious modernization programme in the group, in 1988–89 transferred 25 per cent of the defence procurement funds to other central Government priorities. Nevertheless there are indications that they will proceed with their armament programmes, although not at advertised speed.[8]

While the DDIs face the problem of limited financial resources, the successful completion of the Conventional Armed Forces in Europe (CFE) Treaty could have an indirect effect on the procurement process in the DDIs. Main battle tanks, artillery, armoured combat vehicles and aircraft on the Central European front will be removed. The major NATO countries are considering the possibility of transferring excess hardware to the southern allies, while trying to make sure that advanced systems produced nationally will survive the cuts.[9] Turkey has been involved in such discussions on future transfers with the USA, expressing a firm preference for advanced weapon systems and not outdated equipment. The 1990 eight-year agreement between Greece and the USA regarding the US bases included the transfer of used equipment.[10] The concomitant negotiations with the USA for the procurement of 20 more F-16 fighter aircraft has been closely connected with similar prospects. However, in the near future the implementation of such practices could increase the amount of used equipment transfers to the DDIs, and the final decision for parallel large procurement programmes could become even more complicated (second-hand supplies will be used as the most attractive part of compensation packages).[11]

III. The armament sector

The three DDIs share common characteristics in terms of the existing industrial infrastructure, the small scale of manufacturing plants, limited indigenous technological capabilities and marginal involvement in high-technology sectors.[12] The major activity is assembly work. The weak industrial infrastructure imposes serious problems for the military industrialization programmes. The absence of a strong technological base could limit production to the stage of

[8] See BDMI (note 1), country reports.

[9] See 'Ambassador R. James Woolsey: closing in on a CFE treaty' (interview), *Arms Control Today*, vol. 20, no. 3 (Apr. 1990), p. 7; *Jane's Defence Weekly*, 2 Dec. 1989, p. 1214.

[10] See 'Turkish defense: a special report', *Defense News*, vol. 5, no. 20 (14 May 1990), p. 9; *Jane's Defence Weekly*, 20 July 1990, p. 75.

[11] In France there already is an awareness that a cascading trend could increase the US penetration in military exports. See *Financial Times*, 11 Jan. 1990.

[12] For a review of the manufacturing sectors in Greece, Portugal and Turkey, respectively, see Vaitsos, C. and Giannitsis, A., *Technologikos Metasximatismos kai Oikonomiki Anaptixi*, ['Technological Transformation and Economic Development'] (Gutenberg: Athens, 1987) (in Greek); Laboratorio Nacional de Engenharia e Technologia Industrial, *Technological Development in Portuguese Industry*, Report no. CPA/83-01 (LNETI: Lisbon, 1983); Ansal, H.-K., 'Technical change in the Turkish truck manufacturing industry', thesis presented towards doctoral degree, University of Sussex, 1988.

final assembly; expansion into the production of new military products therefore requires new investments.

Throughout the 1970s and 1980s military manufacturing capabilities have expanded in all three countries. In Greece and Turkey the regional arms race and the search for increasing national independence provided the main rationale for the development of military industry. In Portugal the already operating military manufacturing sector was preserved and expanded moderately.

The development of indigenous military industry was a process of import-substitution.[13] Manufacturing plants were built up whenever a big order could create a sufficient market for a new production line. Some investments were made in established sub-sectors such as ammunition and transportation. From the mid-1980s the large-scale modernization of the Greek and Turkish air forces was accompanied by a shift towards advanced manufacturing in aircraft assembly and passive electronic components.

Companies use military technology transfers as an instrument to secure access to foreign markets. In the case of the DDIs their participation in NATO and the Western alliance's Coordinating Committee on Export Controls (COCOM) favoured a transfer of military technology that could take place without restrictions and export controls. Having this favourable regime and considering the new cycle of procurement expenditures, the major European countries tried to increase their market shares through the promotion of technological and production co-operation. Long-term partnerships, often in the form of joint ventures, were limited to the manufacturing of vehicles and engines for military and civilian end-users. All the major European allies, and lately also Spain, followed this strategy. In the long run the West German involvement proved to be the most successful in terms of export promotion. The key element in this was a military aid programme to finance, for instance, shipbuilding programmes, ammunition plants, tank modernization packages and maintenance plants.[14] The option to expand domestic manufacturing in every contract made the West German offers especially attractive to decision makers in the DDIs.

After the first import-substitution phase, the established arms industries increased the complexity of procurement decisions and presented new dilemmas. In the period under consideration, two trends emerged. Many arms producing companies were involved in 'grey' military exports, and benefited from the Iraq–Iran War (this accounts in part for the increases in exports in the mid-1980s indicated in table 14.2). This export-led growth gave them a strong

[13] See Ayres, R., 'Arms production as a form of import-substitution: the Turkish case', *World Development*, vol. 11, no. 9 (Sep. 1983), pp. 813–23; Albrecht, U., Ernst, D., Lock, P. and Wulf, H., 'Militarization, arms transfers and arms production in peripheral countries', *Journal of Peace Research*, vol. 12, no. 3 (1975), pp. 195–212.

[14] For the interrelations of the German Defence Aid programme with military exports, see Huebner, B., 'The importance of arms exports and armament cooperation for the West German defence industrial base', ed. D. Haglund, *The Defence Industrial Base and the West* (Routledge: London, 1989), pp. 135–36; Timmerman, M., 'German–Turkish cooperation in the area of defence', *NATO's Sixteen Nations*, vol. 32, no. 4 (July 1987), pp. 60–67; *Militärpolitik Dokumentation*, vol. 12, no. 68/69 (1989), pp. 99–101.

Table 14.2. Value of Greek, Portuguese and Turkish arms exports, 1977–87
Figures are in US$ m., current prices.

Country	1977	1978	1979	1980	1981	1982	1983	1984	1985	1986	1987
Greece	..	5	5	..	..	120	..	100	..	40	40
Portugal	60	70	20	50	30	110	60	110	210	220	50
Turkey	..	10	10	10	10	20	90	160	100	..	10

Source: US Arms Control and Disarmament Agency, *World Military Expenditures and Arms Transfers* (US Government Printing Office: Washington, DC, 1989), table II.

impetus for further expansion in military manufacturing projects. Also in each country, politicians and industrialists formed alliances in order to press for follow-on procurements, which in turn led to the expansion of production activities and new investment programmes. Offsets negotiated with the arms suppliers from the larger NATO countries became the main instrument for this policy. Imports of military equipment were expected to create side-effects through the linked importation of technology and negotiated buy-back orders from foreign companies.

The Turkish Government announced an 'export-oriented' policy, pursued personally by Prime Minister, later President, Turgut Özal. The main aim was to create joint ventures with foreign companies in every category of major weapon systems.[15] In Greece the state-owned companies played a prime role in the industry's effort to expand in advanced manufacturing,[16] while in Portugal the arms market was used as a way to secure new orders for existing plants.[17]

Any attempt to examine the armament sector in the DDIs involves confronting an array of difficulties stemming from the lack of previous studies, endemic secrecy and official overestimates of national achievements.[18] Based on information drawn from a large number of open sources, the total work-force of the arms industries of these three countries seems to be in excess of 60 000. Technology imports have been concentrated in the areas of heavy and precision engineering. A majority of the arms production companies also operate assembly plants and an extended network of military-owned factories for maintenance needs. Indigenous technological capabilities are marginal, including the development of reverse-engineering techniques and the assimilation of imported manufacturing technology. Local arms firms have been

[15] See *Defense News*, 26 Sep. 1988, p. 28; *Jane's Defence Weekly*, 7 Nov. 1987, p. 1033; *Jane's Defence Weekly*, 14 Nov. 1987, p. 1109.

[16] See Bartzokas (note 1), chapter 4.

[17] See Luria, R., 'Portugal's armed forces: facing the 1990s', *International Defence Review*, vol. 23, no. 3 (Mar. 1990), p. 279; 'Defence and economics in Portugal', *NATO's Sixteen Nations*, vol. 32, no. 5 (Aug. 1987), pp. 94–98.

[18] See Karasapan, O., 'Turkey's armaments industries', *Middle East Report*, no. 144 (Jan.–Feb. 1987), p. 28; Bartzokas (note 1), chapter 5. For an optimistic view, see Henze, P., *Turkey, the Alliance and the Middle East: Problems and Opportunities in Historical Perspective*, Working Papers no. 36 (International Security Studies Program: Washington, DC, 1981), p. 18.

Table 14.3. Arms production activities in Greece, late 1980s

Area of production	Company	Ownership	Capability
Tanks and military vehicles	ELVO	State	A, D
	Four military factories	Army	A, B, D
	Small private subcontractors	. .	D, E
Aerospace	HAI	State	A, B, D, E
	202nd Air Force Factory	Air Force	B, D
	203rd Air Force Factory	Air Force	B
Shipbuilding	Hellenic Shipyards	State	B, C, D
	Elefsis Shipyard	State	B, C, D
	Salamina Naval Yard	Navy	B
	Souda Naval Yard	Navy	B
Electronic equipment	Alpha	State	A, D, F[a]
	HAI	State	D, F
	306th Military Factory	Army	B, D
	320th Military Factory	Army	B
	Small private companies	. .	A, D, E
Arms, artillery and ammunition	AEEP-K	State	A, D, F
	EVO	State	A, C, E, F
	Small private companies	. .	C, D, E

Abbreviations: A = Assembly line; B = Maintenance and retrofits; C = One-off production; D = Occasional production of metallic parts and/or electronic components; E = Batch production; F = Indigenous product development (reverse engineering).

[a] Recently announced project.

Source: Bartzokas, A., 'Military technology transfer and domestic arms production: the case of Greece', thesis presented towards doctoral degree, University of Sussex, 1991, chapter 4.

successful in exporting military consumables and small arms to the Third World. Some occasional export achievements in the areas of vehicles, patrol boats and medium-calibre fire-arms are due to the special relationships that Portugal and Turkey have established with countries in Africa and the Middle East. Military production is dominated by companies almost exclusively involved in military manufacturing, commonly accounting for more than 70 per cent of total annual sales even in sectors like electronics and trucks. This description fits not only state arsenals but also small, private manufacturers. In the case of Greece, for example, a ring of precision manufacturing shops has been developed that relies heavily on military orders. The same has occurred in Turkey in the area of passive component manufacturers. The activities of the arms industries of the three DDIs are summarized in tables 14.3–14.5.

Table 14.4. Arms production activities in Portugal, late 1980s

Area of production	Company	Ownership	Capability
Tanks and military vehicles	UMM	Private	A, D
	Bravia–Sarl	Private	A, D
Aerospace	OGMA	Air Force	A, B, E, F[a]
	DGFMA	Air Force	B, D
Shipbuilding	Arsenal do Alfeite	Navy	B, D
Electronic equipment	Centrel	Private	A, D
	STE	Private	D
	EID	Private	D
Arms, artillery and ammunition	INEP	State	A, D, F
	Small private companies	. .	C, D

Abbreviations: A = Assembly line; B = Maintenance and retrofits; C = One-off production; D = Occasional production of metallic parts and/or electronic components; E = Batch production; F = Indigenous product development (reverse engineering).

[a] Recently announced project.

Sources: BDMI, *The European Defence Dossier* (BDMI: Southampton, 1990) vol. 3: *Portugal*; Salvador, F., 'Details on the Portuguese frigate programme', *Military Technology*, no. 3 (Mar. 1987), p. 114; Hewish, M., 'Portugal's equipment priorities', *International Defence Review*, vol. 20, no. 12 (1987), pp. 1663–64; *Milavnews*, Feb. 1989, p. 19; 'Defence and economics in Portugal', *NATO's Sixteen Nations*, vol. 32, no. 5 (special issue, 1987).

IV. Future trends

In Greece major manufacturing projects are under way in aerospace, transportation and shipbuilding, although some of these face serious technical and managerial problems. The armament sector has accumulated huge financial debts, and the policy of industrial modernization through the expansion of the military industry has been abandoned. In the near future, the transfer of used equipment from the Central European front in the framework of an agreement resulting from the Conference on Security and Co-operation in Europe (CSCE) process could create additional problems. The Greek Government faces serious economic difficulties and in 1990 asked some of the main state military companies to convert their activities to the production of civil goods. Meanwhile the privatization programme of public enterprises created some interest among European and US firms to buy parts of the Greek armament sector, especially aerospace production facilities, artillery and small arms manufacturing plants and shipyards. The most likely scenario is that the Government will try to preserve the existing military manufacturing infrastructure through subsidies and new defence/public procurement orders. The move towards mergers with foreign interests or even a long-term programme for the conver-

Table 14.5. Arms production activities in Turkey, late 1980s

Area of production	Company	Ownership	Capability
Tanks and military vehicles	MTU	Joint venture	A, D[a]
	OTOKAR	Joint venture	A, D
	OTOMARSAN	Joint venture	A, D
	MANAS	Joint venture	A, D
	MKE	State	D
	Nurol's new project	Joint venture	A,[a] D[a]
	8 maintenance centres at Kayseri	Army	B, D
	Kayseri Army factory	Army	B, D
Aerospace	TUSAS Engines	Joint venture	A, E
	TUSAS-TAI	Joint venture	A, E, F[a]
	ENKA (helicopters)	Joint venture	A,[a] E[a]
	Rokestan	Joint venture	E[a]
	ASELSAN	State	C
	Kayseri Air Force factory	Air Force	B, D
	Other bids for the helicopter project		A,[a] E[a]
Shipbuilding	Taskizak Naval Yard	Navy	B, D
	Golcuk Naval Yard	Navy	B, C, D
Electronic equipment	ASELSAN	State	A, C, D, E
	TELETAS	Private	A, D
	TESTAS	Private	D, E[a]
	ISBIR-ELECTRIC	Armed Forces	A, D
	Kavala plant	Joint venture	A,[a] E[a]
	Marconi-Istanbul	Joint venture	A,[a] E[a]
	Makine Sanayi	Private	
Arms, artillery and ammunition	MKE	State	A, D, E, F
	Small private manufacturers	. .	D

Abbreviations: A = Assembly line; B = Maintenance and retrofits; C = One-off production; D = Occasional production of metallic parts and/or electronic components; E = Batch production; F = Indigenous product development (reverse engineering).

[a] Recently announced project.

Sources: 'Defence and economics in Turkey, *NATO's Sixteen Nations*, vol. 31 (special issue, 1986); Karasapan, O., 'Turkey's armaments industries', *Middle East Report*, no. 144 (Jan.–Feb. 1987), pp. 27–31; 'Turkish defense: a special report', *Defense News*, vol. 5, no. 20 (14 May 1990), p. 9; *Jane's Defence Weekly*, 26 May 1990, p. 1048; *Military Technology*, no. 10 (1987), p. 107; *Financial Times*, 25 Jan. 1989; *Jane's Defence Weekly*, 7 Nov. 1987, p. 1033; *International Defence Review*, vol. 23, no. 2 (Feb. 1990), p. 207; *Defense News*, 13 Nov. 1989, p. 1; *Defence Industry Digest*, Feb. 1990, p. 13; *NATO's Sixteen Nations*, vol. 34, no. 8 (Jan. 1990), market supplement, pp. 34–38.

sion to civilian production could face fierce political opposition and hostility from the armed forces.

Since the mid-1980s, the Government strategy in Turkey has been to create a large-scale modernization programme in order to persuade Western companies to invest. Through drawn-out bargaining, Turkish negotiators succeeded in receiving substantial benefits in terms of cost and extent of technology transfer.[19] The initial strategy was prepared by the Defence Industry Development Agency (DIDA) under close supervision of President Özal. It had two main objectives. The first was to privatize the Turkish arms industry through take-overs and joint ventures, with foreign firms invited to place bids; the second was to increase substantially exports of advanced weapon systems to Middle East customers.[20] The plan was only partly successful, as in the cases of aircraft and engine maintenance and assembly. Other planned projects were slow to come about, for various reasons. Resources were inadequate to carry out procurement programmes and huge investments at the same time. There was also opposition in the military establishment to the creation of an independent agency to co-ordinate new investments. The planned participation of local firms raised yet another series of problems.[21] The prospects for sales of advanced equipment to Arab countries were overestimated, and the buy-back orders from Western firms were limited.[22]

As in the 1970s, Turkish efforts have recently been concentrated in the aerospace sector.[23] It remains to be seen what resources are really available for the implementation of all the planned programmes, and whether the outcome will be different from the current level of 'technology for market access'.

Compared with ambitions in Greece and Turkey, the Portuguese Government has limited ambitions for the domestic arms industry. It has begun a few projects in the areas of ammunition production and military electronics, but the emphasis is on modernizing the armed forces inventory through arms imports.

[19] For complaints by foreign companies, see 'Turkish defense: a special report' (note 10), p. 13.

[20] See *International Defence Review*, vol. 23, no. 2 (Feb. 1990), p. 207; Karasapan (note 18), p. 28. An initial export agreement was signed with Egypt in 1984. See *International Defence Review*, vo. 17, no. 10 (Oct. 1984), p. 1548. The possibility of collaboration was also discussed with Kuwait, Saudi Arabia, the United Arab Emirates and Pakistan. See *Jane's Defence Weekly*, 7 Nov. 1987, p. 1022; and *Jane's Defence Weekly*, 14 Nov. 1987, p. 1109.

[21] See *Country Report: Turkey*, no. 2 (Economist Intelligence Unit: London, 1989), p. 17. The bargaining process linked to a major contract for the local production of combat helicopters is a good example of conflicts. Three groups of companies, including the state-owned TUSAS, are competing in close co-operation with foreign companies for the same contract. See *Jane's Defence Weekly*, 28 July 1990, p. 129.

[22] According to C. Azer, political adviser to the Turkish Defence Minister, while Turkey in 1988 imported military equipment from the IEPG countries to a total value of $1.14 billion, it exported equipment to a value of only $83 895. See 'Turkish defense: a special report' (note 10), pp. 17, 21.

[23] The Turkish Aerospace Industry (TAI) is an assembly plant for F-16s established, at a cost of $4.5 billion, as part of an offset arrangement for the sale of 160 F-16s to Turkey. Future expansion plans include the helicopter project (note 21; see *Flight International*, 13 May 1989, p. 7); local assembly of jet trainers (see *Jane's Defence Weekly*, 22 July 1989, p. 116); new investments in composites manufacturing (see *Interavia*, vol. 45, no. 3 (Mar. 1990), p. 261); a locally designed drone (see *Defence Daily*, 26 Oct. 1989); and participation in the future European Large Aircraft (see *Defense News*, 16 Apr. 1990), p. 6.

V. Conclusion

The three DDIs have developed military industries in an environment of weak industrial infrastructures. Now two of them, Greece and Turkey, are pursuing further expansion in more advanced military projects.

The effects of more intense West European armaments integration are difficult to assess. The hierarchical division of roles inside NATO and the existing technological gap leave little space for DDI military industries. Moreover, in the near future stagnating procurement expenditures could increase protection of national defence markets. Potential suppliers from the DDIs, mainly in the area of metal parts and electronic components, will find it hard to compete in those markets.

The current reconstruction of the West European armament process will have a significant influence on national procurement practices. Pressure for the adoption of 'buy European' clauses and for compensation agreements using surplus equipment will gain in importance. The creation of subcontracting networks will remain a marginal activity, and joint ventures will be designed to satisfy the demands of the armed forces in the DDIs. The DDIs will remain a significant export market for the advanced West European companies. In the cases of Greece and Turkey, future developments will be shaped more by the regional situation than by East–West *détente*. Ironically, for all three DDIs the indigenous military industry has already become an influential factor tipping the scale towards a sustained increase in national procurement expenditures.

15. Sweden's search for military technology

Björn Hagelin

I. Introduction

Political and military analysts from NATO and the EC countries have, at different times, expressed scepticism about the future role and position of non-aligned countries in a Europe in which military production is based on the rational and cost-effective sharing of goods, technology and services. Such scepticism sometimes reflects a lack of understanding of actual non-aligned policy. More specifically, it indicates a lack of awareness of the changes taking place in the neutral countries, not least Sweden, as regards political and military attitudes towards international military co-operation.

Sweden's current policy of national independence and high self-sufficiency in advanced military production was formulated during World War II and strengthened in the course of the cold war period. Two important aspects of this policy should be noted. First, officially stated policy does not always reflect actually implemented policy. Second, the policy was not meant to be flexible but was designed to emphasize consistency and continuity in foreign and defence policies.

Although domestic production does not cover all types of military technology, most types of major conventional weapons and ammunition are or have been manufactured in Sweden, including surface ships and conventionally powered submarines, combat aircraft, tanks and other armoured vehicles, heavy artillery and guided missiles. Despite relatively stable defence expenditures in constant prices from the late 1960s, with shrinking allocations for procurement, there has been strong political and military support of continued acquisition of advanced military equipment. However, while the major military producers in Western Europe can, as NATO members, opt for close 'arm in arm' production, Sweden cannot. Sweden's security policy of remaining 'at an arm's length' has reduced the number of acceptable options when looking for ways out of the dilemma of how to avoid financial and technical constraints while pursuing policies of political independence and high equipment self-sufficiency.

This chapter discusses these aspects of Swedish policy as they relate to Sweden's search for advanced military technology and to future developments in the EC. For the purpose of this discussion, 'relevant technology' is defined as technology necessary for the realization, through indigenous production, of advanced military acquisition programmes. Relevant technology relates here mainly to the production of military equipment. 'Technology' here means goods and know-how in any form needed to implement these programmes. Three questions are addressed: How is relevant technology defined and located? When not indigenous, how and from where is it obtained? How can it be afforded? The answers to all of these questions are important with regard to Sweden's relations with EC countries as well as with other major powers.

II. How to obtain technology

Locating relevant technology

Much basic and applied military R&D takes place inside Sweden. To a certain extent, therefore, relevant technology is defined and created in Sweden. There is one major military research organization, the National Defence Research Institute (FOA), but most military development—as opposed to research—is carried out by industry. In addition, universities, technological institutes and some other educational and basic research institutions also perform some work relevant to military R&D.

If all types of relevant technology were indigenous, the Swedish dilemma would mainly be a financial one. This is not the case. Often cited to illustrate Sweden's degree of self-sufficiency is the claim that imports account for 30 per cent of Sweden's military procurement, that is to say that Sweden is self-sufficient to 70 per cent. This is a distortion of the truth. The percentage is based on the value of total Swedish acquisitions minus an estimated share of foreign imports.[1] During the 1970s and 1980s, Swedish producers increased the share of imported subsystems and components in military equipment produced in Sweden.[2] Even more important, the percentage does not reflect the importance of foreign skills. Although most conventional equipment is manufactured in Sweden, much is based—even dependent—on foreign know-how and technology. This is clearly the case for jet engines, missiles and electronics. All Swedish fighter aircraft have engines manufactured on British or US licences. Previous attempts by Swedish industry to develop air-to-air missiles failed for economic and/or technological reasons, and heavy ground-to-air missiles have been purchased from the UK and the USA. Likewise, all helicopters are imported, as are most light aircraft. Much of the electronics used in Sweden are either manufactured on foreign licences or purchased mainly from the Federal Republic of Germany, Japan and the USA.[3]

Foreign technology becomes known through what might be called the 'military technology intelligence organization'. This includes the Joint Chiefs of Staff, the National Defence Research Institute, the Defence *Matériel*

[1] The share is estimated since data on the size and value of items imported directly by the Swedish defence industry has not been available.

[2] Swedish National Defence Research Institute (FOA), *Svensk försvarsindustri*, report no. C10200-M6 (FOA: Stockholm, Feb. 1982), p. 61.

[3] Swedish Ministry for Foreign Affairs, *Sveriges utrikeshandel i krislägen*, SOU 1989–107 (Government Printer: Stockholm, 1989), pp. 69–71.

Administration and the technical and defence attachés posted abroad. Information about new and potential technologies and weapon systems is gained from printed sources, exchanges of personnel and technology, participation in meetings and conferences as well as through visits to foreign military establishments, installations, industries and exhibitions.

Since Sweden has the ambition to acquire advanced military equipment, the focus is on 'front-edge' rather than old technology. When deployed, the equipment should be advanced and preferably unique (permitting a 'Swedish profile'). The demands on the military technology intelligence organization to find relevant technology abroad are therefore high, despite foreign security restrictions. How is this achieved and who are the main suppliers?

Sweden's policy of international co-operation

Since the failure to create a neutral Scandinavian Defence Pact, and Norway's and Denmark's subsequent applications for NATO membership in 1949, Sweden has not attempted defence policy co-operation—as distinct from military–industrial co-operation—with another country. It was officially stated in the Swedish Parliament on 22 March 1950 that such co-operation would compromise Sweden's neutrality.[4] With the exception of supplying troops and supplies to the UN peace-keeping forces, the Swedish armed forces are committed only to the defence of the national territory.

Swedish participation in military–industrial co-operation has so far involved bilateral government agreements with a few countries. This option has been used restrictively. A 1976 report stated that the economic rationale for co-operation (i.e., cost-savings) is 'usually' incompatible with indigenous military production.[5] As expressed in the report, co-operation in Western Europe for the production and acquisition of military supplies has moved towards closer integration, which presupposes co-ordination of national defence, economic and security policies. In short, developments towards closer military–industrial co-operation in Europe are here seen as a parallel to closer political co-operation. In view of Sweden's policy of independence, it was regarded as undesirable to take formal part in such multinational co-operation.

However, official declarations and reviews dealing with the implementation of such a restrictive policy have been far from clear. Why did the option of closer military–industrial co-operation appear at all on the Swedish political agenda? In 1968 Defence Minister Sven Andersson declared that Sweden might be forced to leave its traditional policy of indigenous and independent production for economic as well as technical reasons: the development towards

[4] See the statements by the Prime Minister, before the First Chamber of Parliament, and by the Foreign Minister, before the Second Chamber, in *Documents on Swedish Foreign Policy 1950–1951* (Swedish Ministry for Foreign Affairs: Stockholm, 1957), pp. 13–26.

[5] Swedish Ministry of Defence, *Säkerhetspolitik och totalförsvar*, SOU 1976–5 (Government Printer: Stockholm, 1976).

larger economic markets (i.e., the EC) and the fast and cost-increasing developments in military technology.[6]

Not mentioned was a third reason that became apparent in the later handling of the issue. Swedish participation in military–industrial co-operation was discussed in a 1971 Government bill concerning foreign military sales.[7] Foreign military sales and international military–industrial co-operation were, and still are, seen as possible and closely related solutions to the Swedish armament dilemma. In other words, from at least the early 1970s the economic and technical necessity to co-operate with other countries in military projects was officially recognized.

The involvement of Swedish firms in international military–industrial co-operation is not subject to the same degree of Government control as are war *matériel* sales. A number of official investigations carried out in the late 1980s lend support to industry demands for international co-operation in suggesting that the existing, relatively relaxed control requirements for such co-operation be maintained. It has been acknowledged officially[8] that military industries, like other industries, need to participate in technical co-operation,[9] and that if Sweden were not involved in international co-operation it would be unable to sustain military production within important industry sectors.[10]

The use of the word 'usually' in the 1976 report indicated a willingness to accept exceptions from the formally restrictive policy. In the 1977 security policy bill only 'obligatory co-operation' was excluded.[11] What constitutes such co-operation is clearly a matter of political interpretation, and the term has subsequently never been defined. In fact, the 1968 statement by Defence Minister Andersson was not merely a hypothetical reflection but reflected changes that were already being implemented. The first Memorandum of Understanding (MOU) regarding 'the exchange of secret technical information of a military nature' was signed with the US Government in 1952.[12] This arrangement made it possible for Sweden to buy military equipment and technology directly from US Government agencies through the US Foreign Military Sales programme and not only, as before, through commercial channels.

[6] Andersson, S., 'Försvarets betydelse för svenskt näringsliv', *Vårt Ekonomiska Läge* (Svenska Sparbanksföreningen: Stockholm, 1968).

[7] *Kungl. Maj:ts proposition angående krigsmaterielexport*, Government bill no. 146, 1970/71 (Government Printer: Stockholm, 1971).

[8] Swedish Parliament, *Totalförsvarets fortsatta utveckling*, Defence Committee Report no. 11, 1986/87 (Swedish Parliament Printer: Stockholm, 1987).

[9] Swedish Ministry of Defence, *Det svenska totalförsvaret under 90-talet*, SOU 1987–9 (Government Printer: Stockholm, 1987), p. 105.

[10] Swedish Ministry for Foreign Affairs, *Utlandssamverkan på krigsmaterielområdet*, SOU 1989–102 (Government Printer: Stockholm, 1989).

[11] Swedish Ministry of Defence, *Inriktningen av säkerhetspolitiken och totalförsvarets fortsatta utveckling*, Government bill no. 74, 1976/77 (Government Printer: Stockholm, 1977).

[12] *Agreement between the United States of America and Sweden, Mutual Defense Assistance; Reciprocal Procurement Rights for Military Supplies or Services*. For the text of the Agreement, see *Treaties and Other International Acts, Series 2480* (US Department of State: Washington, DC, 1952), or *Sveriges överenskommelser med främmande makter, 1954*, no. 74 (Government Printer: Stockholm, 1954). The Agreement has since been modified.

Like later agreements signed with the governments of Norway, Denmark and Canada, the Swedish–US MOU also made common procurement possible. This permits Sweden to place orders for a given type of equipment concomitantly with, for instance, the US Marine Corps, thereby reducing delivery time. The price can also be assumed to be lower than if Sweden were to place the order as an individual foreign customer. MOUs also smooth the way for a constant input of foreign military equipment, know-how and services to Sweden. Most likely, bureaucratic paperwork as well as the first stages of political, commercial and military 'red tape' can also be avoided. As of 1991 Sweden was party to over 15 such bilateral MOUs, signed with all major West European countries, Canada, the USA and some other nations.

In view of the restrictions that generally apply to military and 'strategic' sales, why is Sweden not, as a non-aligned nation, excluded from sharing US and NATO advanced technology? In part it is due to Sweden's long tradition of democracy and anti-communism. After 1945 social democracy became the model which, together with a relatively high electoral threshold for parliamentary representation, effectively neutralized communist parties. Swedish traditions and history express, in general, a favourable attitude towards Western values and ideas. Moreover, the USSR has traditionally been regarded as the most likely potential aggressor.

A factor of particular relevance is Sweden's strategic location in northern Europe. Intelligence about the USSR has been of primary political and military importance to decision-makers not only in Sweden but also in the USA and other NATO countries. The short distance across the Baltic Sea has given Sweden a good forward location for gathering such information. The co-operation and sharing of intelligence information with US and other Western intelligence agencies seem to have given Sweden access to certain advanced and restricted technology.[13]

However, in order to be eligible to receive advanced NATO technology, Sweden has also had to accept NATO rules and regulations. Although not a formal member of the Coordinating Committee (COCOM), until 1967 Sweden informally followed the COCOM embargo policy towards communist states for militarily relevant civilian (dual) equipment and technology.[14] After 1967 the US Department of Commerce granted sales of such items to Sweden on receipt of 'Letters of Assurance', in which importing companies assured in writing that they would inform the US suppliers, and in the last instance the US Government, about the use of the purchased technology. The Swedish companies also assured that they would fully co-operate with US authorities. Several Swedish

[13] Hagelin, B., 'Nordic armaments and military dependencies', *Current Research on Peace and Violence*, vol. 9, nos 1–2 (1986), pp. 13–27; Gummesson, J. and Kanger, T., 'Sveriges bluff om neutraliteten', *Aftonbladet*, 26 Oct. 1990; Gummesson, J., 'ÖBs hemliga avtal med USAs militär', *Aftonbladet*, 27 Oct. 1990; Gummesson, J. and Kanger, T., 'Sanningen om den svenska neutraliteten', *Aftonbladet*, 3 Dec. 1990; Cole, P. M., 'USA-samarbetet hemlighölls', *Dagens Nyheter*, 21 Oct. 1990; Cole, P. M., 'Neutraliteten gavs upp', *Dagens Nyheter*, 31 Oct. 1990; Nilsson, B., 'Vi spelade dubbelt', *Dagens Nyheter*, 6 Nov. 1990; Åström, S. 'Undén ingen Mr. Hyde', *Dagens Nyheter*, 7 Nov. 1990.

[14] Hagelin, B., *Kulorna rullar* (Ordfront förlag: Stockholm, 1985), pp. 19–32.

companies signed such Letters of Assurance in order to acquire advanced US technology.[15]

Sweden was, in fact, the object of discrimination. The US Government only requested Letters of Assurance in connection with sales to Sweden. The Swedish Government was eventually willing to accept further US demands in order to end this discrimination. In 1986, as the last of the European neutrals to do so, Sweden passed a law regulating the re-transfer of advanced technology produced in Western countries.[16] A primary reason for taking this step was to secure access to such technology.[17] Finally, in 1990 the Swedish civilian control mechanisms were expanded to include the issuing of import certificates, by which Swedish importers promise to observe US export regulations. As a result the US Department of Commerce deleted Sweden from its civilian (dual) control list.[18] The only other country for which Swedish import controls are used is South Africa.[19]

The MOUs are instruments for regulating trade in advanced technology. From what is known, the importance of the MOUs varies. It appears that implementation in the past has been confined less to the exchange of R&D than to the manufacture and/or direct purchase of foreign military equipment. However, the more recent agreements may point towards a new path marked by R&D co-operation with some of the larger West European states, Canada and the USA. *Ad hoc* Swedish co-operation in international projects has been projected to increase as a result of increasing involvement in co-development.[20]

In certain respects the JAS 39 Gripen combat aircraft is an illustration of this development. It is the first major Swedish military project in which the contracted firms were advised by the Government to search actively for foreign suppliers in order to keep development costs down. Apart from the engine, electronics and other foreign subsystems and components, British Aerospace (BAe) under a co-operative arrangement supplied Saab, one of the prime contractors, with new lightweight material know-how for construction of the wings. BAe also designed, developed and manufactured the first wing series. Saab established an office in Great Britain for designs that could not be made in Sweden due to lack of skilled designers. There was also co-operation between Ericsson and Ferranti for the JAS 39 Gripen radar (partly identical technology is used in Ferranti's Sea Vixen radar developed for the Sea Harrier aircraft).

The increasing internationalization of Swedish military production is shown by several indicators: the increase in joint ventures, in Swedish firms' shares in foreign companies, in the establishment of Swedish companies abroad and in

[15] Holmström, M. and von Sivers, T., *Tekniken som vapen* (Ingenjörsförlaget: Stockholm, 1985), pp. 44–45.

[16] Swedish Ministry for Foreign Affairs, *Press Release*, 27 Feb. 1986.

[17] Swedish Ministry for Foreign Affairs (note 3), p. 71.

[18] Holmström, M., 'USA häver restriktioner för teknikhandel', *Svenska Dagbladet*, 12 Sep. 1990.

[19] Hagelin, B., 'Arms transfer limitations: the case of Sweden', ed. T. Ohlson, SIPRI, *Arms Transfer Limitations and Third World Security* (Oxford University Press: Oxford, 1988), pp. 165–66.

[20] Swedish Ministry for Foreign Affairs, *Försvarsindustrins utlandsverksamhet*, SOU 1987–8 (Government Printer: Stockholm, 1987).

Table 15.1. Number of new Swedish military joint ventures, 1950–85 and 1986–89

Equipment category	1980	1981	1982	1983	1984	1985	1950–85	1986–89
Weapons/ammunition	–	4	2	2	15	5	32	38
Vehicles	–	1	–	1	3	1	6	9
Electronics	–	–	1	1	1	1	4	9
Total	–	**5**	**3**	**4**	**19**	**7**	**42**	**56**

Sources: Swedish Ministry for Foreign Affairs, *Försvarsindustrins utlandsverksamhet*, SOU 1987–8 (Government Printer: Stockholm, 1987); Swedish Ministry for Foreign Affairs, *Utlandssamverkan på krigsmaterielområdet*, SOU 1989–102 (Government Printer: Stockholm, 1989).

imports of foreign components for Swedish projects. The number of new military–industrial joint ventures in 1986–89 was higher than the number for the whole period 1950–86 (see table 15.1). The extent of Swedish involvement in individual projects has also increased. The majority of companies involved are those developing and manufacturing arms and ammunition (in contrast to those producing vehicles, electronics or other military supplies). Again, most of the partners are in Western Europe, mainly in the FRG and the UK, as well as in the USA. As specified by the companies, the main purpose is to increase markets. Lack of know-how on the part of the Swedish firm and specific demands on the part of the purchaser(s), are also important motivating factors—more than 50 per cent of production in joint ventures was said to be the result of foreign demands.[21]

A new war *matériel* sales bill is expected to be passed in the course of 1991. Due to its controversial content, it will most likely be presented after the parliamentary elections in September. The bill will focus on measures to control Swedish involvement in international projects, mainly in co-development. A policy that would put further restrictions on arms sales and international co-operation should not be expected. A strong and broad common interest exists among influential policy makers, Government agencies and groups on the one hand, and the military–industrial leadership and employees on the other—a 'military–industrial–political brotherhood'—in support of sustained military production. The result has been an 'armament race', not with an explicit and well-defined enemy, but with military technology itself. Despite financial difficulties, the Swedish Government seems prepared to support quality and mobility at the expense of quantity for Sweden's future defence.[22]

Thus, there is no automatic link between Swedish armament behaviour and political changes in Europe. In order to be affordable in the short term, this race is more and more assumed to have to include industrial co-operation and military sales—in 1988 over half of Sweden's war *matériel* production was exported. As an illustration of the new situation, a Conservative Party member

[21] Swedish Ministry for Foreign Affairs (note 20).
[22] 'Sverige skrotar invasionsförsvaret', *Svenska Dagbladet*, 4 Mar. 1991.

of the Defence Committee suggested in September 1990 that Sweden should be permitted to export war *matériel* to the new democracies in Eastern Europe.[23] Also, in December 1990 former Defence Committee Chairman, Carl Johan Åberg, suggested that one or more foreign partners in the JAS 39 Gripen project would reduce Sweden's financial burden and increase the aircraft's potential foreign market.[24] Such discussions seem already to be taking place.[25] The Hägglund & Söner company is also negotiating co-production of the next Swedish light tank with companies in several European countries. If these negotiations are successful a large part of the production will probably be located abroad.[26]

While the official investigations discussed above do not question the economic and technological rationales for international co-operation, they do argue for 'sufficient' controls and for safeguarding Swedish security interests. However, it seems clear that the security rationale for assuring Swedish military self-sufficiency, hitherto of top priority, has become one of secondary importance. The perspective seems rather to be to sustain, for unclear technological and commercial reasons (in which prestige perhaps also plays a role), an advanced indigenous military industry. In 1971, when most of the present basic export control guidelines were formulated, it was considered impossible for Sweden to become a partner in a multinational military project. The proposed guidelines point in a new direction, as do economic realities.

III. How to afford advanced military production

For a small country, Sweden has a relatively large and advanced military industrial base, with production in a broad range of weapon categories. Six of Sweden's major producers are among Europe's top 100 arms companies.[27] Four of them are among the world's top 100 arms companies.[28] The cost of sustaining this industrial base is substantial in terms of money, skills and technology. The military accounts for over 20 per cent of the nation's R&D.[29] Of the roughly 25 000–30 000 people (measured in man-years) employed in the arms industry, one third is employed by Bofors and by Saab's aircraft division. Plants in Linköping, Karlskoga and Stockholm account for more than half of the employment in Sweden's arms industry.[30]

[23] 'Moderat vill sälja vapen till öst', *Upsala Nya Tidning*, 10 Sep. 1990.

[24] Åberg, C. J., 'Rädda Jas med utländsk hjälp!', *Dagens Nyheter*, 6 Dec. 1990.

[25] Östlund, B., 'Fransk jakt på nya partner', *Svenska Dagbladet*, 9 July 1990.

[26] 'Hägglunds på väg mot miljardavtal', *Svenska Dagbladet*, 12 Mar. 1991.

[27] 'Europe's top 100 defence companies', *Defence & Foreign Affairs*, vol. 18, nos 1–2 (Jan./Feb. 1990), pp. 18–20; see also appendix A.

[28] Anthony, I., Courades Allebeck, A., Gullikstad, E., Hagmeyer-Gaverus, G., Miggiano, A. and Wulf, H., 'Arms production', SIPRI, *SIPRI Yearbook 1991: World Armaments and Disarmament* (Oxford University Press: Oxford, 1991), appendix 8A, pp. 310–316.

[29] Swedish National Bureau of Statistics (SCB), *Forskningsstatistik*, report PM 1991–01–25 (SCB: Stockholm, 1991), table 7.

[30] Swedish Ministry for Foreign Affairs, *En politik för nedrustning och utveckling*, report no. Ds1988:19 (Swedish Ministry for Foreign Affairs: Stockholm, 1988).

There are, theoretically, ways out of, or at least ways to reduce, the effects of the 'vicious' acquisition circle described in chapter 1 of this volume. However, few of these can be implemented without impinging on the established goals, discussed above, of national independence and military self-sufficiency. To increase military expenditure to an extent that would solve the problems has been politically unacceptable for a long time. The critical financial situation has forced the Government to move forward the five-year defence decision from 1992, as scheduled, to 1991. The debate surrounding the decision is plagued by political controversy. The Defence Committee decided in October 1990 not to present a detailed report.[31] Instead, Supreme Commander Bengt Gustafsson was requested to present new information for the 1991 decision. The political guidelines presented in 1991 signify an important shift in military doctrine. Instead of maintaining that the major threat is an invasion preceded by a period of warning that would permit general mobilization, the new doctrine envisages a core of well-equipped units prepared to meet a variety of situations, including a surprise attack. Outlays associated with increased reliance on high-cost quality equipment are to be offset by reduced personnel and administrative costs, to be achieved through reductions in standing forces and the peacetime organization.[32]

An indigenous military industry is important to the Government's guidelines. Costs are to be reduced or maintained at current levels through industry rationalizations, and through increases in imports, exports and international co-operation. Military production is already concentrated in a relatively small number of major producers. There is one aircraft producer (Saab), one producer of heavy army equipment (Bofors), one producer of submarines (Kockums) and one producer of surface ships (Karlskronavarvet). For certain types of equipment, production is concentrated in two companies. This is the case for guided missiles (Saab and Bofors), and light army vehicles (Bofors and Hägglund & Söner). In both instances some joint production occurs.

The Government has acknowledged that in order to be competitive, the national military–industrial base must be reduced and become more specialized. In a report presented in 1990 the Supreme Commander defined his priorities and criteria for the future Swedish arms industry.[33] In terms of equipment areas, he stressed the need for continued Swedish capacity to produce ammunition, electronics and torpedoes. Indigenous capacity should also be sustained in technologies that are protected abroad for security reasons. In general, however, the report foresaw reduced indigenous military development of major military systems, further reductions of the industrial base towards fewer producers,

[31] *5 rapporter från 1988 års försvarskommitté*, SOU no. 108 (Swedish Ministry of Defence: Stockholm, 1990). This Defence Committee publication contains reports from five working groups and is not to be considered a Committee report.

[32] 'Sverige skrotar invasionsförsvaret', *Svenska Dagbladet*, 4 Mar. 1991.

[33] Överbefälhavaren [Supreme Commander], 'Försvarsindustrin', supplement no. 15 to *Försvarsmakt 90* (Försvarsstabens informationssektion: Stockholm, 1990).

Table 15.2. Foreign share of industry financing of Swedish military R&D, 1977–89

Outlay figures are in m. Swedish crowns, at current prices.

		Financed by industry			
				Foreign sources	
Year	Total outlay[a]	Outlay	Percentage	Outlay	Percentage
1977	604	108	*18*	22	*4*
1979	579	140	*24*	19	*3*
1981	796	254	*32*	23	*3*
1983	1 181	596	*51*	24	*2*
1985	2 233	875	*39*	65	*3*
1987	2 652	996	*38*	242	*9*
1989	2 596	969	*37*	229	*9*

[a] The share of outlay not financed by industry is paid by Swedish military organizations.

Source: Swedish National Bureau of Statistics, Stockholm (data supplied to author on request).

more production under international co-operation arrangements and more acquisitions through imports.

Industrial mergers and the selling and purchasing of peripheral ventures on the domestic military market were important activities in the 1980s. The result has been further specialization, in particular within the electronics sector. Bofors Electronics has become a major shareholder of Ericsson Radar Electronics, making Bofors one of Sweden's major military electronics firms. Nobel Industrier and FFV have together established Swedish Ordnance–FFV/Bofors AB, which consists of Nobel Industrier's ammunition and powder production facilities, including the Bofors company, and FFV Ordnance.[34] In a joint venture with FFV Aerotech, Volvo Flygmotor AB has become majority owner of Volvo Aero Support AB, specializing in aircraft engine overhauls.[35] Kockums Marine became owner of Karlskronavarvet in late 1990.

The main rationale for increased internationalization of military R&D, production and acquisition is to share costs. As table 15.2 shows, throughout the 1980s there was an increase in the share of foreign financing of Swedish military R&D. During the same period, Swedish producers gained a stronger foothold within the NATO and EC military markets by way of MOUs, shared company ownerships and joint ventures. One Swedish firm with multinational involvement in military R&D and manufacturing is Ericsson—by holding a majority share in Italian FIAR, for instance, Ericsson is via the Ferranti consortium involved in the European Fighter Aircraft radar contract. The USA has also become a more important market for Swedish products.[36] Examples are the Bofors/Ford Aerospace's Divisional Air Defense (DIVAD) vehicle

[34] 'Flera hundra jobb är i farozonen', *Upsala Nya Tidning*, 28 Dec. 1990.

[35] 'FFV Aerotech bildar ett nytt bolag med Volvo Flygmotor', *FFV-Nytt*, nos 11–12 (1990), pp. 10–13.

[36] Tusa, F., 'Among NATO nations, Swedish equipment is "right stuff"', *Armed Forces Journal International*, vol. 127, no. 11 (June 1990), p. 24.

(cancelled due to technical difficulties with the radar), FFV's Ranger Anti-armour, Anti-personnel Weapon System (RAAWS) and all-terrain vehicles from Hägglund & Söner.

As noted above, an essential criterion for Swedish participation in joint ventures is foreign demand. Since most MOUs and other forms of international military co-operation are with nations in Western Europe and with the USA, developments in and among the European nations, as well as relations across the Atlantic, are important for Sweden's future options.

IV. Sweden and the West European arms industry

Figure 15.1 is an attempt to illustrate, in a simple manner, four policy options for future Swedish military–industrial co-operation with Western Europe. They are, by necessity, very speculative and dependent on a variety of factors. The options are defined by West European military–industrial relations on the one hand, and by US–West European relations on the other. Option 2 is close to today's military structure, with US military involvement in Western Europe but with no solution to the pull and push between co-operation and competition in West European military production. Option 1 is the development of a 'European pillar' within the NATO structure. West European military integration may also take place without the participation of the USA. This is the EC option 3. However, were such a far-reaching political change to take place, there might also be room for a planned reduction of European armaments. Another alternative is that the pull and push will continue even after a US withdrawal. Both these alternatives are within the bounds of option 4.

While the Organization for Economic Co-operation and Development (OECD) and the European Free Trade Association (EFTA) were perceived and presented by Swedish Government representatives as purely economic frameworks without political constraints, the EC was regarded as a potential political and security grouping. The single market to be achieved in 1992 and the decision to steer the EC towards political union in 1993 have sharpened the dilemma for Sweden between what is perceived as economically necessary and what is politically desirable. Ireland is already a member of the EC, and Austria has applied for membership. In the course of 1990 the Swedish Government changed its position from rejecting to promoting EC membership.

After a series of articles in the major dailies, Prime Minister Ingvar Carlssson entered the media debate in May 1990 by stating that the transformation of the EC into a political union would make it impossible for Sweden to become a full member.[37] He later specified two factors that would make Swedish membership possible: the elimination of the risk of a European war and the scrapping of any plans of a co-ordinated EC foreign or security policy.[38] The statement created a heated political debate and resulted in a special meeting of the parliamentary

[37] Carlsson, I., 'EG-medlemsskap omöjliggörs', *Dagens Nyheter*, 27 May 1990.
[38] 'Politisk strid om EG-medlemsskap', *Upsala Nya Tidning*, 30 May 1990.

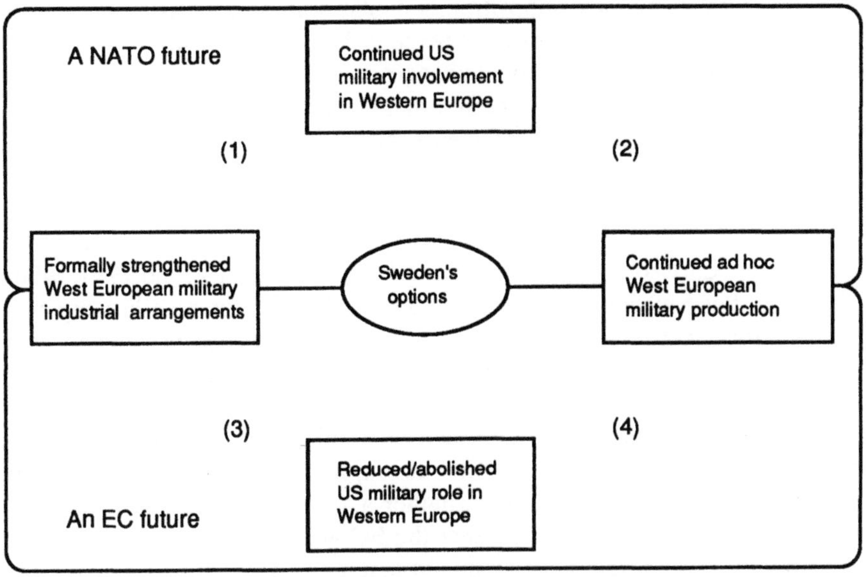

Figure 15.1. Sweden's policy options for future military–industrial co-operation with Western Europe: 'at an arm's length' or 'arm in arm'

Foreign Affairs Committee. When the Prime Minister again published his views on the issue in July, the distinction between short- and long-term developments was much more pronounced, and a political union was no longer perceived as a definite hindrance to Swedish membership. However, the Prime Minister maintained that Sweden's freedom of action would increase if, instead of the EC plans for foreign, security and defence policy co-operation, a pan-European security arrangement could be realized.[39] This view was supported by Foreign Minister Sten Andersson in September 1990, just a few days before the Social Democratic Party congress. In November 1990 a majority of the parliamentary Foreign Affairs Committee voted in favour of a Swedish membership application.[40] On 12 December, barely six months after the Prime Minister's article in May, the Parliament decided that Swedish membership is compatible with neutrality.[41] The Government was given a 'green light' to apply for membership when it saw fit. Although the Government has not considered it a necessary pre-condition, it has already accepted a fundamental law change which permits Swedish courts to receive foreign assistance in the interpretation of laws and regulations.[42] Moreover, Sweden's largest translation project began during 1990 with the translation to Swedish of the EC's legal source material.

Many factors contributed to this sudden shift in government policy. The December decision came as a response to a variety of pressures, including the

[39] Carlsson, I., 'EG-hinder kan undanröjas', *Dagens Nyheter*, 5 July 1990.
[40] 'Majoritet bakom svensk EG-ansökan', *Upsala Nya Tidning*, 23 Nov. 1990.
[41] 'Majoritet röstade för EG', *Upsala Nya Tidning*, 13 Dec. 1990.
[42] 'Grundlagen ändras för EG', *Upsala Nya Tidning*, 7 Dec. 1990.

attitude of the parliamentary opposition, a public opinion increasingly positive to Swedish EC membership, a growing perceived 'threat' of being left 'outside' after 1992, and the dramatic political changes that swept Eastern Europe in the autumn of 1990. In addition, economic difficulties undermined many of the arguments pointing at the risks of membership. In a declaration on foreign policy presented in February 1991, Foreign Minister Andersson stated that 'When the world around us goes through dramatic changes, the conditions for our peacetime foreign policy also change'.[43] This conclusion was repeated by Prime Minister Carlsson in his statement to Parliament on 14 June 1991, when the decision that Sweden would submit an application for membership was officially announced.[44] However, EC membership may also be viewed as a logical step in the 20-year economic struggle to save (parts of) Sweden's military industry.

Although an application was submitted on 1 July 1991, it will take several years for the EC's decision to come through. Much can happen in the meantime—it is too early to know whether a future pan-European security arrangement can be realized, or whether the EC will refrain from military and security policy co-operation, as Prime Minister Carlsson hoped. As for the latter, debates within the EC certainly include discussion of a co-ordinated defence structure, possibly involving the Western European Union (WEU) or, as suggested by EC Commission Chairman Jacques Delors in March 1991, the establishment of EC multinational and intervention forces.[45]

In a speech on 14 June Prime Minister Carlsson stated that 'Sweden cannot take part in a common defence policy or a mutual defence commitment within the EC framework without foregoing its policy of neutrality'.[46] However, to reject co-operation 'within the EC framework' is not the same as excluding such co-operation with EC members. In fact, the Prime Minister said in the same speech that the desire to strengthen a West European 'defence policy profile' is likely to lead to increased security and defence policy co-operation within the WEU: 'It is likely that some form of permanent relationship will be established between the EC and the WEU'. This would thus not consititue co-operation 'within the EC framework', and the Prime Minister did not explicitly exclude Swedish participation in WEU activities.

Assuming that a political union is not seen as a problem for Sweden's application for EC membership, these discussions may lead to one of two things. The first is that the EC could co-ordinate its military/security policy *before* Sweden became a member. Sweden would then remain free to abstain from participating, by choosing not to become a member. Foreign Minister

[43] 'Regeringens utrikespolitiska deklaration', *Från Riksdag & Departement*, no. 9 (1991), p. 13.
[44] Press release (Prime Minister's Office: Stockholm, 14 June 1991), unofficial translation.
[45] 'Västeuropeisk samordning', *Upsala Nya Tidning*, 22 Aug. 1990; 'EG på väg bli militärallians', *Upsala Nya Tidning*, 9 Mar. 1991.
[46] Press release (note 44).

Andersson has in fact stated that if the EC decides in favour of a military union, Sweden will not accept membership.[47]

The alternative, namely, that a co-ordinated EC military/security policy is established *after* Sweden's application has been granted is more problematic. There has been no official attempt to explain the impact this would have on Sweden's policy choices. The official Swedish position is that this is a very unlikely alternative. As consensus within the EC would be needed to establish a military union, the argument runs, an EC defence policy would 'in all likelihood' not be realized.[48] The conclusions of the European Council meeting in Rome on 14–15 December 1990 to maintain ties to NATO and to honour member countries' defence obligations[49] support the Swedish position.

It has been stated in official contexts that Sweden's foreign policy should not be based on probabilities. However, it could well be maintained that this is precisely what Sweden is doing by sustaining a military defence, or by basing its EC policy on the calculated unlikeliness of a coherent EC military policy. On the other hand, to doubt the latter may in retrospect show the same lack of foresight as did those who in the 1940s doubted the possibility of Franco-German co-operation, or those who in May 1990 rejected the possibility of Sweden applying for membership in the EC.

Were the EC to pursue an integrated defence policy after Sweden has become a member, as suggested above, Sweden would have the difficult choice of either remaining a member, at the expense of its military non-alignment, or leaving the EC. In the first case, Sweden's policy of neutrality may of course be redefined, or perhaps even abandoned. Should Sweden instead want to leave, the primarily economic difficulties associated with non-membership would remain unresolved. To exit is one thing, but what is there to enter?

In the early plans for a West European military acquisition organization presented in 1978, the door for non-NATO members' participation in military co-operation was left open through the Independent European Planning Group (IEPG).[50] From an EC point of view, it would not be efficient to close the door on Sweden's military–industrial capacity—and, as noted above, Sweden is not against closer military–industrial co-operation. It is important to realize that Sweden can, and most likely will, increase its participation in international military projects whether or not any formal changes take place in its foreign or defence policies.

Nevertheless, formally and extensively linking West European security and military integration is contrary to Swedish short-term policy. While politics may be 'the art of the possible', each possible option has a price. In formulating future guidelines and general policy for multinational military co-operation, the

[47] 'Inget svenskt medlemsskap om EG skaffar armé', *Upsala Nya Tidning*, 12 Mar. 1991.

[48] 'Svenskt EG-beslut redan i vår?', *Upsala Nya Tidning*, 6 Dec. 1990.

[49] European Council, *European Council, Rome, 14 and 15 December 1990, Presidency Conclusions*, internal report, no. SN 424/1/90 (OR. f) Rev 1.

[50] European Parliament, *European Parliament Working Documents 1978–1979*, document no. 83/78 (European Parliament: Luxembourg, 1978), Annex 1.

Swedish Government will be obliged to declare what price it is prepared to pay—the political price of walking 'arm in arm' with other EC countries in military–industrial co-operation, or the economic and technological one of remaining 'at an arm's length'. Although increasing industrial participation in multinational military projects will most likely be politically accepted from 1992, a certain hesitation will likely also prevail against giving the industry totally free hands in arranging such co-operation. The political dilemma will not disappear with only a change in the military sales policy.

As West European threat perceptions are reassessed in light of the changes that have taken place in the former Warsaw Treaty Organization (WTO) countries, Western arms producers may give new emphasis to export drives aimed at non-European countries. Given the decline in the global arms market during the 1980s, increased West European military co-operation may therefore lead to efforts to reduce foreign market competition by harmonizing military export policies. Such attempts may leave smaller suppliers without much of a foreign military market, or may incorporate them in solutions, defined by the larger producers, in which production is oriented to suit foreign customers. Efforts to support production in European countries with less developed military industries may serve such purposes.

If Sweden accepts participation in multinational military projects it may have to give up important aspects of its restrictive war *matériel* sales policy. This issue has been the object of scrutiny in two recent official investigations. Although both suggest that Sweden should not give up its restrictive policy, one concluded that it may, in some cases, be necessary to do so in order to receive foreign orders. According to the study, such exceptions may be permitted when Swedish supplies are included in the final product only to a minor extent or do not make a difference to the product's foreign components, or when there are strong Swedish defence policy interests for co-operation, and the partner(s) demand or wish to re-export the final product.[51] As has been described above, the latter's interests and demands are already clearly visible. Quite understandably, in Prime Minister Carlsson's speech on 14 June 1991 the EC Intergovernmental Conference guidelines regarding economic and technological co-operation in arms production and the formulation of a common arms export policy were only mentioned in passing.[52]

One factor working in Sweden's favour is that the pace of West European military–industrial co-operation has been slow. The process has been characterized by *ad hoc* co-operation and step-by-step solutions. Despite far-reaching plans and ambitions, the ambiguities concerning West European integration remain. This gives the Swedish Government a certain freedom of action, or rather the freedom not to change current policy. The CFE Agreement, signed in Paris in November 1990, does not entail total European disarmament. Soviet military action in the Baltic republics in the first months of 1991 show that the

[51] Swedish Ministry for Foreign Affairs (note 10), p. 154. Similar arguments are voiced in Ministry of Defence (note 31).

[52] Press release (note 44).

use of military force remains an option in Europe. The Chief of the Swedish Air Force has pointed out that Soviet offensive systems withdrawn from the CFE central zone are to be deployed in the Baltic and Leningrad regions,[53] while Defence Minister Roine Carlsson stated in late November 1990 that nothing has yet reduced the importance of Sweden's air defence.[54] The Supreme Commander wrote in 1991 that the changes in Europe do not motivate a lower ambition for Sweden's defence.[55] The Prime Minister also said on 14 June 1991 that the concrete efforts to reduce tensions in Europe are not as manifest in Northern Europe as they are in Central Europe. Security policy planning must always assume that the unexpected may occur, he said. New tensions may develop and war cannot be categorically ruled out: 'This is why we maintain our policy of neutrality'.[56]

If West European foreign policy and defence integration pick up pace, and if this takes place at the expense of transatlantic co-operation, Sweden's position will be further complicated. There are both political and economic reasons for this. The official Swedish position is that as long as NATO remains, there is no need for the EC to develop into a military organization. Sweden will therefore support NATO as the European military organization until a pan-European security solution is reached. Swedish and US ambitions are thus for the time being in harmony.

In the short term a pragmatic Swedish policy is supported by the fact that it is becoming more and more difficult to distinguish between civilian and military technologies. The Single European Act is not irrelevant for the future European arms industry, since it will affect arms companies through their civilian activities. Swedish participation in EC industrial programmes aims at strengthening Sweden's European identity, to increase its research efficiency and to prepare Swedish industry for the European market.[57] One area for such co-operation is R&D,[58] most importantly in high-technology areas such as telecommunications, information and aerospace technology. Sweden participates in Eureka, the EC R&D and technical co-operation programme, with approximately 50 projects. By becoming a member of the EC, Sweden can both influence the direction of R&D in Europe and benefit from EC project support, perhaps also in the military sector.

Even with the change of government in 1991, Sweden will already have applied for membership, and may become a member by 1995. The impetus will be to continue the step-by-step process of moving away from the Swedish policy of association 'at an arm's length', by engaging in EC dual-use high-

[53] 'Utländskt köp blir dyrare', *Upsala Nya Tidning*, 10 Dec. 1990.

[54] 'JAS-projektet fortsätter på rimliga villkor', *Upsala Nya Tidning*, 1 Dec. 1990.

[55] Överbefälhavaren [Supreme Commander], *Brief Summary ÖB 92: Material Compiled by the Supreme Commander of the Swedish Armed Forces for the 1992 Resolution on Defence* (Försvarsstabens informationssektion: Stockholm, 1991).

[56] Press release (note 44).

[57] *Sverige–EFTA–EG* (Swedish Ministry for Foreign Affairs: Stockholm, 1989), p. 105.

[58] *Del av Europa: Sverige och den västeuropeiska integrationen* (Swedish Ministry for Foreign Affairs: Stockholm, 1988).

technology projects and in multinational military projects through the IEPG and the European Co-operative Long-term Initiative for Defence (EUCLID), a European version of the US Defense Advanced Research Project Agency (DARPA). Closer co-operation will also involve research projects at universities and institutes of technology. This again introduces a controversial issue first raised in the context of Swedish–US co-operation in R&D during the Viet Nam War, namely the relationship between basic research and military applications. The debates of the late 1960s and early 1970s may reappear, and focus on the EC as sponsor of militarily relevant basic and exploratory research projects.

In the course of these developments, Sweden will lose its distinct policy and equipment profile. By seeking to integrate its military production and acquisition apparatus with those of the EC countries, the Swedish Government has clearly shown that it now regards isolated independence as a worse alternative than integrated dependence. Sweden is entering a third period in the development of its post-war foreign and defence policy. The first two lasted roughly 20 years each. The first, from 1949 to 1968, was marked by a not yet fully developed policy of non-alignment. A national nuclear weapons option was kept open, and foreign support was regarded as possible in case of attack. The second period, from 1968 to 1991, and was summarized in the slogan 'non-alignment in peacetime aiming at neutrality in wartime'. Its central attributes were political independence and a high degree of self-sufficiency in conventional arms production and military preparedness. This is quickly becoming history as a third period begins, propelled by the economic necessity for international co-operation, and the recognition that past levels of independence no longer are affordable. The survival of Sweden's military industry will no longer depend only on national priorities, but will increasingly depend also on international developments and the state of the global military market. By the year 2010 foreign military sales and international co-operation will have become instruments of Sweden's foreign policy. Then it may again be time for Sweden to make an important policy choice.

Part IV
Summary

16. Summary

Michael Brzoska and Peter Lock

I. Introduction

Reactions to the fundamental changes in the framework of arms production in Western Europe, outlined in the introduction and described in more detail in the chapters of part II by Ian Anthony and Herbert Wulf, Harald Bauer, Frederic Pearson, Judith Reppy, and Philip Gummett and William Walker, have differed from country to country, as reported in detail in the country analyses in part III. This is not surprising, given the many influences on arms production. Arms production links security issues and economic policies, the prerogatives of private producers and a core of state activity. Policy is made by a large number of actors—such as various state governments, industrial interest groups and unions, armed forces and a large number of commercial enterprises—representing a range of interests in a number of competing areas of concern, including foreign and alliance policy, military affairs, disarmament, technology, regional politics, employment and profits.

The diversity in perceptions and actions, in policies and politics, has repercussions both for the political effort to create a more cohesive structure of arms production and for the analytical effort to predict the future of the industry in Western Europe. This is one reason why the questions asked in the introduction could only partly be answered by the other contributors to this study. The framework of arms production is still under construction in several respects, both as regards the future military demand for weapons and the political landscape of Europe. The strength of the analyses lies rather in describing what happened in the recent past than in predicting what will happen in the future.

Section II of this summary highlights some of the points made in the country analyses. Section III discusses common issues and open problems, building on the thematic chapters of part II of the book.

II. The country analyses

During the 1980s the British Government was at the forefront of efforts to introduce more competitive elements into the arms procurement market. While these focused on the domestic market, Prime Minister Margaret Thatcher's Government also pressed for a more competitive West European arms market. The British initiatives were clearly linked with the overall drive for a liberalization of markets, including the privatization of most state-owned enterprises.

Trevor Taylor argues that the British reforms were successful. What used to be a static industry, with comparatively low productivity, became a trim, in most sectors highly competitive, industry. As a result, arms production is one of the few sectors of British industry that now thrives on world markets. Still, quite a number of arms producing companies lost business, and workers jobs, in the process.

Privatization and the more competitive environment hastened the concentration process. British Aerospace became the dominant British arms producer, except in electronics, where GEC grew substantially larger. This concentration, which partly offset the efforts to create more competition, increased the incentive for the Government, as well as the companies concerned, to push for a wider market, or rather markets. In the British case, both the Government and the industry are interested in the wider West European market, as well as in the US market where British firms were rather successful in selling products in the 1980s. This two-way approach is conflict-prone, as was exemplified by the clash between various ministers about the rescue of the helicopter producer Westland.[1] Of course, this issue is linked to the larger issue of the role the USA is to play in the future West European security framework.

In France restructuring has remained limited, suggesting that 'value for money' played a less important role in arms procurement than in the UK. French protectionism with regard to the national arms industry has many roots. Some are linked to the establishment of the centralized French state, with its strong military component, and others are connected to the efforts, begun in the 1960s, to modernize industry. As Claude Serfati shows, arms production is run by the state, or rather by a close network of generals, bureaucrats and managers under the auspices of the Ministry of Defence.

Although the costs of this protectionist policy increased significantly in the 1980s, large arms exports allowed a profitable combination of 'welfare' and 'warfare'.[2] An important cause of the subsequent crisis was the fall in Third World arms imports. In addition, France was left managing the debt-burden of its best customer in the early 1980s; Iraq. As a result, the financing of arms production sectors which had been held afloat by exports, such as shipbuilding and tank production, became a problem in the late 1980s. Also, earlier assumptions about the spin-off benefits of arms production to other industries were increasingly questioned.[3] Jacques Fontanel and Louis Pilandon discuss how even a successful company such as Thomson had to resort to large-scale government financial support.

[1] At that time, the group—supported by Prime Minister Margaret Thatcher—favouring a sale to whoever offered the most attractive terms (even if it was a US firm), won over the group—led by then Defence Minister Michael Heseltine—seeking a West European solution. See, for example, Freedman, L., 'The case of Westland and the bias to Europe', *International Affairs*, vol. 64, no. 1 (1987), pp. 1–19.

[2] See Kolodziej, E. J., *The Making and Marketing of Arms* (Princeton University Press: Princeton, N.J., 1987).

[3] François Chesnais emphasized this point during the conference from which this book originated. For a more detailed discussion see Chesnais, F. (ed.), *Compétitivité internationale et dépenses militaires* (Economica: Paris, 1990).

On one level, French interests in a larger West European arms market are identical to those in the UK. In pursuit of widening markets, French companies have been very active in buying arms producing companies in other parts of Western Europe. On another level, pertaining to government involvement in industry, the French approach is markedly different. Government ownership, state economic support of industry and local government, and the use of arms production as an instrument of regional and technology policies will remain a central component of French armaments policy for some time. Of course, this approach depends on the maintenance of an independent French security policy.

In the Federal Republic of Germany, concentration in the arms market has come about without much pressure from increased competition. Traditionally, the Government has co-operated closely with arms producers. While it has sought to decrease procurement costs, efforts to open the market have largely been frustrated by effective political lobbying on the part of company managers.

This largely inefficient state of affairs might have continued had not the automobile producer Daimler Benz, under its Chairman Edzard Reuter, moved to restructure the company into an all-round high-technology enterprise. By taking over AEG, Dornier, MTU and MBB, Daimler Benz concentrated a large share of German arms production under its roof. At the time of these take-overs, Daimler Benz was careful to secure guarantees for government orders for a few years. However, it also must now confront the shrinking national as well as international arms markets. The company's response has been to seek international alliances, not only in Western Europe, but also in the USA and in Japan. For the company's management, a more integrated West European arms market is a step in the direction of becoming a major actor in a competitive world market for armaments and other high-technology goods.

Some smaller German arms producers have already achieved such competitive positions, although only in market niches. Werner Voß discusses shipbuilding companies in northern Germany that have full order-books despite the decline in both the German and international markets. One element in their success is that they have managed to integrate components from a large number of sources.[4]

The Italian arms industry largely escaped restructuring efforts in the 1980s, despite a severe reduction in exports and domestic demand. Giulio Perani and Mario Pianta show the diversified structure which enhanced flexibility during the growth period in the 1970s and early 1980s but turned the comparatively small Italian arms industry into a national economic problem in the late 1980s and early 1990s. Although a large part of the industry is state-owned, there is no centralized leadership as in the French case. By way of example, there are strong rivalries both between and within the state-owned companies IRI and EFIM, as in the dispute within IRI over the integration of Aeritalia and Selenia into the new company Alenia. Although the competition by larger companies

[4] This type of integration of parts was thoroughly analysed in Faltas, S., *Warships and the World Market* (Martinus Nijhoff: The Hague, 1986).

from France, the FRG and the UK constitutes a threat to Italian arms producers, government and industry are now in favour of a more integrated West European arms market, in the hope of overcoming the current crisis.

In Sweden, restructuring of arms production occurred at a fast pace in the late 1980s, resulting in a high degree of concentration in one company under joint state and private control. Björn Hagelin shows how this process was shaped by anticipating future competition from a more integrated West European market. The Swedish arms industry is set to survive, despite its comparatively small size.

Restructuring of arms production has been quite dramatic in some of the smaller, more industrialized West European countries. In the Netherlands, the market leader, Philips, sold its extensive arms production interests, including sizeable subsidiary companies in several West European countries and in the USA. The main buyer was Thomson of France. The traditional Belgian arms producer, Fabrique Nationale de Herstal, was rescued from bankruptcy by the French arms manufacturer GIAT. In Switzerland, Oerlikon-Bührle—at one time among the more prosperous West European arms producers—is taking losses, and has put its arms production divisions up for sale. These cases demonstrate how difficult it has become for arms industries not supported by a large national market, or a generous government, to survive in the more competitive environment of the late 1980s and early 1990s .

In the less industrialized countries of Southern Europe, governments are willing to subsidize arms production because they thereby expect to be able to introduce technologies that then can be spread to other sectors of their national industrial bases. They are therefore in favour of greater integration of West European arms production and a lowering of barriers to technology flow. They do not fear the more intense competition that comes with integration, as they do not expect to be subject to the full force of the market. The larger West European producers have already agreed to give special treatment to producers in Greece, Portugal and Turkey. Also the Spanish industry, as described in detail by Jordi Molas-Gallart, seems isolated from full competition. Moreover, for all these countries, a more integrated West European market is not the only alternative. Each has strong links with US companies via collaborative projects.

III. The issues

The special treatment of the less industrialized countries of Southern Europe, as proposed in the Vredeling Report and agreed to by the Independent European Programme Group (IEPG),[5] violates the principles of free competition that should be the core of a more integrated West European arms market. While probably necessary to secure the support of these states for a more open market, this commitment opens a Pandora's box of other claims of exception from the principle of the open market.

[5] See chapter 3, pp. 40–42.

Everywhere in Western Europe small industries are threatened by increased competition. Will they be protected? Government backing in France and Italy gives companies in these countries an unfair advantage over companies in the UK and the FRG. Will it be tolerated? The large government military research and development (R&D) outlays in France and the UK give companies in these countries a head start over companies in countries with smaller R&D budgets. How can this be levelled out?

One approach often used in the past was to collaborate in the development and production of weapon systems. On paper, the importance of collaboration is overwhelming. According to statistics collected by Simon Webb, countries in Western Europe spend between one-fourth (France and the UK) and three-fourths (Belgium and the FRG) of their procurement expenditures on joint projects.[6] However, because of careful application of the principle of 'juste retour', very little money is spent outside the respective countries. In the mid-1980s the Vredeling Commission still judged joint development and production as the most likely compromise between national protectionism and open markets. However, scepticism about the benefits of co-operation has since increased; several projects, including the NATO frigate programme, have been cancelled. Interestingly, as arms producing companies have begun to play a more dynamic role in shaping the framework of procurement decisions, they have partly substituted administered co-production projects by forming long-term transnational working relationships, such as MBB and Aérospatiale in helicopter production, Matra and MBB in drones, and BMW and Rolls Royce in engines.

In addition to the economic issues, the political aspects of arms production raise questions about the probabilities and possibilities of a more integrated West European arms market. The degree of government intervention varies from the strictly liberal UK to mostly state-run France, Greece and Spain. Dependence on foreign military supplies is not an issue in the smaller countries, while an important one in France, which is not part of the NATO military structure. The Vredeling Report clearly comes out in favour of a 'defence industrial policy', the deliberate use of arms production as an instrument of general technology policy. This may partly be due to the fact that the majority of the 'wise men' in the Commission had strong links to the arms industry. It is partly also a reflection of the fear that Europe is losing the race for new technologies to Japan and the USA. This was a greater worry for many Europeans in the early to mid-1980s, when the Report was written, than it seems to be in the early 1990s. The success of the Vredeling Report is closely linked to the widespread European interpretation of the US Strategic Defense Initiative as a technological–industrial measure.

The relationship between the USA and Western European is a crucial issue shaping the future of arms production. Not surprisingly, the US Government and much of the US arms industry are pushing for a NATO-wide open market

[6] Webb, S., *NATO and 1992: Defense Acquisition and Free Markets*, RAND report no. R-3758-FF (RAND Corp.: Santa Monica, Calif., July 1989), p. 20.

for arms.[7] The argument that procurement costs would be much lower is probably as valid as the European concern that only a few West European producers would survive in such a market. However, Western Europe is not alone in seeking to disguise protectionist practices with free market rhetoric. As the US Government promotes a more open arms procurement market, Congress, in particular the House of Representatives, and national security intellectuals fend against the perceived dependence of US arms procurement on overseas purchases.[8]

Political issues are also at the heart of the question of arms exports. The disclosure of West European arms supplies to Iraq has in the wake of the Persian Gulf War strengthened interests in stricter control, but not in a uniform way. The old divisions between rather lax French export practices and comparatively strict laws in the FRG, for instance, seem to be stronger than before. Despite the urgency of this matter, in view of the Single Market envisaged for 1993, no serious negotiation on this issue has become public. This is partly due to unresolved institutional questions. As described by Harald Bauer, rivalry between the IEPG, the Western European Union (WEU) and the European Community (EC) intensified in 1990.[9] A concentration of the arms industrial activities, including common arms export policy, under the roof of the EC seems the most likely outcome. This would leave the IEPG as a body to co-ordinate military planning, with the aim of standardizing equipment and procedures. European Political Co-ordination might increase to become the sole forum for the harmonization of security policy in a broader sense. The WEU might be retained for some time as the body to co-ordinate military policy in a more narrow sense, that is, the deployment and use of armed forces.

IV. The outlook

Many questions remain about the future of arms production in Western Europe. However, there are also near certainties: technology will progress; Europe will integrate more fully; and arms spending will continue to decline.

As Herbert Wulf and Ian Anthony predict, this will lead to large reductions in the work-forces of arms producing companies in Western Europe. Restructuring in the name of economic efficiency will make redundancies even larger than reductions in demand suggest. This is the result of the inefficiency and over-capacity that were allowed to exist in arms production in the past. At the same time, it is not likely that economic forces alone will dictate the

[7] The main US spokesperson on his issue has been the US ambassador to NATO, William Taft. See for example Taft, W., 'Standing together: defense security and industrial cooperation in the new NATO', *NATO's Sixteen Nations*, vol. 35, no. 5 (1990), pp. 14–19.

[8] For an overview see Moran, T. H., 'The globalization of America's defense industries', *International Security*, vol. 15, no. 1 (1990), pp. 57–90. See also US Congress, Office of Technology Assessment, *Arming Our Allies: Cooperation and Competition in Defence Technology* (US Government Printing Office: Washington, DC, 1990).

[9] See for instance Sir Geoffrey Finsburg (rapporteur), *Report submitted on behalf of the Political Committee of the Assembly of Western European Union: WEU and the European Community*, Document no. 1250 (Western European Union: Paris, 14 Nov. 1990).

changes in the arms industry. National interests and concern about specific policies, such as those pertaining to technology, regional development and employment, will influence this change in a significant way.

For example, collaborative projects are likely to continue. Although economically inefficient, they allow governments to use procurement policy as an instrument in other policy areas. Likewise, protectionist policies will be continued in a number of countries. The costs of letting the market do all the restructuring are high, in terms of losses in employment, technology and national pride. It is difficult to imagine that governments will not pursue arms industrial policies to counter these effects. Conversion, the transformation of arms production capacity to civilian production, while not highly valued by West European governments, has a role here, as has more general economic support for regions and communities hit by unemployment.[10]

In the end, it is likely that more unity will bring more diversity. States with competitive arms industries, such as the UK, may allow more open markets than states with weak industries. It is more difficult to eliminate national protection at the level of end-products than at the level of components, where it is already low. Predictably, transnationally linked arms producers will increase in number, thus diminishing the national perspective on arms production.

Seen from this perspective, integration in arms production on a West European scale has been going on for some time. Until the mid-1980s, integration in arms production lagged behind integration in comparable industries. It picked up speed with the EC decision, in 1986, to introduce the Single Market, and again after it became clear that there would be widespread reductions in arms spending. After 1988 there was a great increase in the number of mergers and acquisitions (see appendix B). This industrial activity has helped to make the arms industry less of a special case. The next step is to establish an institutional framework for integration. In this process the arms industry is already lobbying for a framework that will serve its interests, while other interest groups are lagging behind. Unfortunately, it is too early to end on a positive note. The 'normalization' process of arms procurement and production is incomplete. What is lacking is institutionalized, parliamentary control over the whole process, including, as a first priority, a common arms export policy.

[10] See the policy discussion in Paukert, L. and Richards, P. (eds), *Defence Expenditure, Industrial Conversion and Local Employment* (International Labour Office: Geneva, 1991).

Appendices

Appendix A. The 50 largest arms-producing companies in Western Europe, 1989

Ian Anthony, Agnès Courades Allebeck, Gerd Hagmeyer-Gaverus, Paolo Miggiano and Herbert Wulf

Table A.1 contains information on the 50 largest arms-producing companies in Western Europe in 1989. Companies with the designation *S* in the column for rank in 1989 are subsidiaries; their arms sales are included in the figure in column 6 for the holding company. Subsidiaries are listed in the position where they would appear if they were independent companies. In order to facilitate comparison with data for the previous year, the rank order and arms sales figures for 1988 are also given. Where new data for 1988 has become available, this information is included in the table; thus the arms sales figures for some companies which appeared in table 8.2 in the *SIPRI Yearbook 1990* and in table 8A in the *SIPRI Yearbook 1991* have been revised.

Sources and methods

Sources of data. The data in the table are based on the following sources: company reports, a questionnaire sent to over 300 companies, and corporation news published in the business sections of newspapers and military journals. Company archives, marketing reports, government publication of prime contracts and country surveys were also consulted. In many cases exact figures were not available, mainly because companies often do not report their arms sales or lump them together with other activities. Estimates were therefore made.

Arms sales. The criterion for the rank order of companies is their arms sales in 1989 (column 6). The arms sales figures are based on the sources mentioned above and thus not comparable to the SIPRI arms transfer figures given in the *SIPRI Yearbook 1991*, chapter 7.

Coverage. The data are for 1989; data in columns 2 and 7 are for 1988. The fiscal year for companies is not always the calendar year. No calculations have been made to adjust fiscal to calendar years.

Exchange-rates. Most figures collected were given in local currencies. To convert figures into US dollars, the period-average of market exchange-rates of the International Monetary Fund, *International Financial Statistics,* was used.

Profit. Profit after taxes is shown for the entire company, not for the arms-producing sector alone. For figures taken from journals and periodicals, it was not always clear whether profit was given before or after taxes.

Employment. The figure shown is either a year-end or yearly average number for the entire company, as published in the sources used.

Key to abbreviations in column 5. A = artillery, Ac = aircraft, El= electronics, Eng = engines, Mi = missiles, MV = military vehicles, SA/O = small arms/ordinance, Sh= ships, and Oth = other

Table A.1. The 50 largest arms-producing companies in Western Europe, 1989[a]

Figures in columns 6, 7, 8 and 10 are in US $ million.

1	2	3	4	5	6	7	8	9	10	11
Rank					Arms sales		Total sales	Col. 6 as % of	Profit	Employment
1989	1988[b]	Company[c]	Country	Industry	1989	1988[d]	1989	col. 8	1989	1989
1	1	British Aerospace	UK	Ac El Mi SA/O	6 300	5 470	14 898	*42*	3 546	125 600
2	2	Thomson S. A.	France	El Mi	4 320	4 470	12 027	*36*	78	100 000
3	4	Daimler Benz	FRG	Ac Eng MV El Sh	4 260	3 420	40 634	*10*	3 622	368 226
S	S	Thomson-CSF (Thomson SA)	France	El Mi	4 120	4 320	5 282	*78*	412	21 723
S	–	DASA (Daimler Benz)	FRG	Ac Eng El Mi	3 930	0	7 484	*53*	– 58	62 959
4	3	Direction des Constructions Navales	France	Sh	3 630	3 580	3 626	*100*	. .	30 460
5	5	GEC	UK	El	2 880	2 970	14 408	*20*	1 118	107 435
6	9	IRI	Italy	Ac Eng El Sh	2 230	1 950	41 285	*5*	1 178	363 449
7	7	Dassault-Breguet	France	Ac	2 200	2 080	3 059	*72*	46	15 572
8	6	Aérospatiale	France	Ac Mi	2 190	2 300	4 969	*44*	32	36 899
S	8	MBB (DASA)	FRG	Ac El Mi	1 840	1 990	3 336	*55*	16	24 194
9	11	CEA Industrie	France	Oth	1 820	1 640	5 254	*35*	345	36 820
10	10	Lucas Industries	UK	Ac	1 640	1 760	3 655	*45*	313	55 957
11	12	EFIM	Italy	Ac MV El	1 510	1 490	3 754	*40*	– 18	38 000
12	15	INI	Spain	Ac A MV El Sh SA/O	1 290	1 290	16 118	*8*	693	149 910
13	12	FIAT	Italy	Eng	1 280	1 530	36 695	*3*	2 410	286 294
14	16	SNECMA Groupe	France	Eng	1 260	1 270	3 371	*37*	46	25 702
15	14	Rolls Royce	UK	Eng	1 220	1 410	4 857	*25*	382	46 000
16	22	Oerlikon-Bührle	Switzerland	Ac A El SA/O	1 040	930	2 903	*36*	– 25	27 326
17	18	GIAT	France	A MV SA/O	1 020	1 150	1 056	*97*	– 107	14 200
S	S	Aeritalia (IRI)	Italy	Ac El	1 020	880	1 860	*55*	41	14 903
18	21	Nobel Industries	Sweden	El Mi SA/O	950	1 010	3 556	*27*	214	22 246

19	25	Siemens	FRG	El	900	800	32 515	*3*	839	365 000
20	19	Matra Groupe	France	Mi Oth	870	1 040	3 462	*25*	95	21 242
21	24	VSEL Consortium	UK	MV Sh	870	830	879	*99*	46	16 610
22	20	Philips	Netherlands	El	800	1 010	26 984	*3*	648	304 800
S	S	MTU (DASA)	FRG	Eng	780	970	1 946	*40*	7	17 654
S	S	Bofors (Nobel industries)	Sweden	A El Mi SA/O	740	810	765	*97*	. .	6 315
S	–	Telefunken Systemtechnik (DASA)	FRG	El	730	0	1 043	*70*	. .	10 779
S	–	Matra Défense (Matra)	France	El	710	0	710	*100*	. .	. .
23	38	Hunting Associated Industries	UK	SA/O	650	440	1 228	*53*	80	. .
24	27	Diehl	FRG	A MV El SA/O	620	610	1 374	*45*	. .	15 300
S	S	Agusta (EFIM)	Italy	Ac	610	490	940	*65*	15	8 426
S	S	Dornier (DASA)	FRG	Ac El Mi	590	570	1 172	*50*	. .	10 247
25	32	Smiths Industries	UK	El	590	530	1 156	*51*	183	13 606
S	S	Oto Melara (EFIM)	Italy	A MV Mi	580	530	580	*100*	8	2 329
26	31	Eidgenössische Rüstungsbetriebe	Switzerland	Ac Eng A SA/O	550	550	583	*94*	. .	4 828
27	28	Thyssen	FRG	MV Sh	540	600	18 218	*3*	439	136 091
28	17	Thorn EMI	UK	El	540	1 200	6 094	*9*	521	65 444
29	30	SAAB-SCANIA	Sweden	Ac Eng	530	570	7 219	*7*	156	48 708
S	S	SNECMA (SNECMA Groupe)	France	Eng	530	770	2 108	*25*	13	13 727
30	34	FFV	Sweden	A El SA/O Oth	510	500	993	*51*	6	9 903
31	26	Rheinmetall	FRG	A SA/O	510	650	1 402	*36*	65	13 700
S	S	Electronique Serge Dassault (Dassault-Breguet)	France	El	500	510	645	*78*	21	4 109
32	35	Racal Electronics	UK	El	490	480	3 237	*15*	233	37 414
S	S	Selenia (IRI)	Italy	El	490	380	649	*76*	11	6 655
33	33	Krupp	FRG	MV El Sh Oth	490	510	9 406	*5*	– 240	63 557
S	S	CASA (INI)	Spain	Ac	480	500	684	*70*	– 51	10 138
34	29	Ferranti-International Signal	UK	El	470	580	1 304	*36*	71	12 700
35	44	SAGEM Groupe	France	El	410	350	1 628	*25*	36	16 459
36	37	Devonport Management	UK	Sh	410	450	418	*98*	. .	7 500
S	S	FIAT Aviazione (FIAT)	Italy	Ac Eng	410	660	684	*60*	8	4 651

1	2	3	4	5	6	7	8	9	10	11
Rank					Arms sales		Total sales	Col. 6 as % of	Profit	Employment
1989	1988[b]	Company[c]	Country	Industry	1989	1988[d]	1989	col. 8	1989	1989
37	39	Dowty Group	UK	Ac El	400	410	1 199	*33*	90	14 680
38	36	Westland Group	UK	Ac	390	450	708	*55*	28	. .
39	45	Renault Véhicules Industriels	France	Eng MV	390	340	27 351	*1*	1 458	174 573
S	–	Siemens Plessey Electronic Systems (Siemens)	UK	El	390	0	. .	. .	. .	. .
S	S	FIAT IVECO (FIAT)	Italy	MV	380	270	3 794	*10*	54	22 969
40	40	Ericsson	Sweden	El	360	390	6 250	*6*	327	69 229
41	–	Mannesmann	FRG	MV	360	0	11 878	*3*	268	125 800
S	43	Krauss-Maffei (Mannesmann)	FRG	MV	360	380	718	*50*	3	5 320
42	41	Hawker Siddeley	UK	El	350	390	3 519	*10*	203	42 600
S	S	Sextant Avionique (Thomson-CSF/Aérospatiale)	France	El	350	0	943	*37*	27	9 800
43	48	Ascom Holding	Switzerland	El	330	330	1 620	*20*	31	15 351
S	S	Fincantieri (IRI)	Italy	Sh	330	310	1 310	*25*	– 185	20 941
S	S	Hollandse Signaalapparaten (Thomson-CSF)	Netherlands	El	330	410	337	*98*	16	4 900
44	61	Lürssen	FRG	Sh	320	170	402	*80*	. .	1 176
S	S	Mercedes Benz (Daimler Benz)	FRG	MV	320	380	29 982	*1*	596	223 219
S	70	Short Brothers (Bombardier)	UK	Ac Mi	310	140	347	*89*	. .	7 053
45	42	Vickers	UK	Eng MV SA/O	300	390	1 214	*25*	65	12 238
S	S	SAAB Aircraft Division (SAAB-SCANIA)	Sweden	Ac	300	380	627	*48*	. .	6 940
S	S	EN Bazan (INI)	Spain	Sh	300	380	372	*81*	5	10 342
S	S	Santa Barbara (INI)	Spain	Sh	300	170	372	*81*	. .	600
46	49	HDW	FRG	Sh	290	310	528	*55*	8	4 520

S	S	SAGEM (SAGEM Groupe)	France		280	350	775	*36*	20	7 243
47	51	Société Nationale des Poudres et Explosifs	France	A SA/O	280	280	580	*48*	7	6 711
S	S	Blohm & Voss (Thyssen)	FRG	MV Sh	270	260	546	*49*	1	5 610
S	S	Krupp Atlas Elektronik (Krupp)	FRG	El	270	290	399	*68*	. .	4 324
S	S	FFV Ordnance (FFV)	Sweden	SA/O	270	270	286	*94*	. .	2 456
48	53	Wegmann	FRG	MV	270	260	530	*51*	. .	4 750
49	46	FN	Belgium	Eng A SA/O	260	340	536	*49*	– 53	6 001
S	S	Contraves Italiana (Oerlikon-Bührle)	Italy	A El	260	0	260	*100*	. .	1 020
S	S	Standard Elektrik Lorenz (Alcatel Alsthom)	FRG	El	260	320	2 142	*12*	30	20 876
50	54	Labinal	France	El	250	230	953	*26*	43	15 320
S	S	LITEF (Litton Industry)	FRG	El	250	0	. .	. .	. .	. .
S	S	Turboméca (Labinal)	France	Eng	250	230	371	*67*	18	3 883

[a] Both the rank designation and the arms sales figures for 1988 are also given, in columns 2 and 7, respectively, for comparison with the data for 1989 in columns 1 and 6.

[b] A dash (–) in this column indicates that the company did not produce arms in 1988, in which case there is a zero (0) in column 7.

[c] Company names in parentheses after the name of the ranked company are the names of the holding companies. The parent companies, with data pertaining to them, appear in their rank order for 1989.

[d] A zero (0) in this column indicates that the company did not produce arms in 1988, but began arms production in 1989, or that in 1988 the company did not exist as it was structured in 1989.

Note: The authors acknowledge the assistance of Bernard Harbor (London), Sami Faltas (Eindhoven), Ernst Gülcher (Antwerp), Peter Hug (Bern), Evamaria Loose-Weintraub (Stockholm), Arcadi Olivares i Boadella (Barcelona), Mario Pianta and Giulio Perani (Rome), Paul Rusman (Haarlem) and Werner Voß (Bremen) in the data collection.

Appendix B. International mergers, formation of new companies and take-overs in the arms production sector of Western Europe, 1989–90

Ian Anthony and Herbert Wulf

Table B.1. International mergers and formation of new companies in the arms production sector of Western Europe, 1989–90

Companies	Country of origin	Name of merged/new company	Year
Dense-Pac Microsystems Inc.	USA	. .	1989
Hybrid Memory Products Ltd	UK		
MTU	FRG	MTU Turbomeca Rolls-Royce	1989
Turbomeca	Italy		
Rolls-Royce	UK		
Sagem	France	Italiana Sistemi Inerziali	1989
Sepa	Italy		
Matra space activities	France	Matra Marconi Espace	1989
GEC Marconi space activities	UK		
Aérospatiale	France	Eurocopter	1990
MBB	FRG		
Matra	France	Eurodrone	1990
MBB	FRG		
Aérospatiale	France	. .	1990
OFEMA	France		
AIROD	Malaysia		
Thomson-CSF	France	Defence Electronics of Singapore	1990
Allied Ordnance	Singapore		
Thomson-CSF	France	Ferranti-Thomson Sonar Systems	1990
Ferranti International sonar division	UK		
BMW	FRG	BMW Rolls Royce	1990
Rolls Royce	UK		
Selenia	Italy	CSEE Defense	1990
CSEE	France		
Santa Barbara (INI)	Spain	DEFTEC	1990
MBB	FRG		
CESELSA	Spain	Aeronautical Systems Designers	1990
SD-Scicon	UK		
Ferranti Technologies	UK	Ferranti Tech Asia	1990
Allied Ordnance Company	Singapore		

Table B.2. International take-overs in the arms production sector of Western Europe, 1989–90

Buyer company/ companies	Head office	Seller company	Head office	Year
Bombardier Inc.	Canada	Short Brothers PLC	UK	1989
Matra	France	Fairchild Space, Fairchild Communications and Electronics, Fairchild Control Systems	USA	1989
SNECMA	France	FN Moteurs	Belgium	1989
Thomson-CSF	France	HSA	Netherlands	1989
Thomson-CSF	France	TRT	Netherlands	1989
Alcatel	France	ACEC Space, Defence and Telecommunications Division	Belgium	1989
Siemens	FRG	Plessey Radar and Defense Systems	UK	1989
Diehl	FRG	BGT	USA[a]	1989
Elsag	Italy	Bailey Controls	USA	1989
Nobel Industrier	Sweden	Philips Elektronik-Industrier	Netherlands	1989
ASA	UK	RJO Enterprises Inc.	USA	1989
Hunting	UK	Irvin Industries	USA	1989
Dowty	UK	Palmer Chenard Industries	USA	1989
Astra	UK	BMARC	Switzerland[b]	1989
SABCA	Belgium	Dassault Belgique Aviation	France/Belgium	1990
Thomson-CSF	France	NV Philips MBLE Defence	Netherlands	1990
Thomson-CSF	France	Link-Miles	UK	1990
Thomson-CSF	France	Ferranti Sonar division	UK	1990
GIAT	France	Fabrique Nationale (and parts of Beretta, Italy)	Belgium	1990
Aérospatiale Alcatel Selenia	France, France, Italy	Parts of Space Systems Loral (previously bought from Ford)	USA	1990
MAN	FRG	Steyr-Daimler-Puch	Austria	1990
Sauer (Susco Acquisition)	FRG	Sundstrand-Sauer Co.	USA	1990
Fincantieri, Bremer Vulcan AG	Italy, FRG	Sulzer Diesel AG	Switzerland	1990
Finmeccanica	Italy	Ferranti Italiana[c]	UK	1990
Finmeccanica	Italy	FIAR (Italy, owned by Ericsson, Sweden)	Sweden	1990
FIAT	Italy	Enesa	Spain	1990

Buyer company/ companies	Head office	Seller company	Head office	Year
Thorn EMI	UK	MEL (British military electronics part of Philips)	Netherlands/UK	1990
Astra Holdings	UK	PRB	Belgium	1990
Royal Ordnance of British Aerospace	UK	Heckler & Koch	FRG	1990
Dowty Group	UK	Resdel Engineering	USA	1990
BEI Electronics	USA	4 divisions of Systron Donner[d] (Inertial, Seaton-Wilson, Edcliff and Duncan Electronics)	UK	1990

[a] Bodenseewerk Gerätetechnik GmbH (BGT) was owned by the US Perkins Elmer Group.

[b] BMARC was a subsidiary of the Swiss Oerlikon Group.

[c] Ferranti Italiana, previously owned by the UK group Ferranti International, controls five companies active in the avionics and communication sector: LABEN, ELMER, OTE, EAE and PRO-EL.

[d] Systron Donner is owned by Thorn-EMI of the UK.

Source: SIPRI arms production data base.

Appendix C. European NATO major weapon procurement expenditure, 1981–90

Figures are in US $m., at constant (1988) prices.

	1981	1982	1983	1984	1985	1986	1987	1988	1989	1990
Belgium	650	611	595	545	518	548	560	492	400	409
Denmark	396	390	386	359	308	301	339	334	296	291
France	7 490	7 878	8 255	8 151	8 492	8 850	9 648	9 496	9 746	9 100
FRG	6 843	6 533	6 811	6 743	6 760	7 082	7 100	6 773	6 652	5 703
Greece	689	583	497	569	535	498	541	790	682	663
Italy	2 469	2 540	2 883	2 778	3 128	3 122	3 954	4 188	4 091	3 024
Luxembourg	1.1	1.5	1.1	1.0	2.5	2.0	2.9	2.4	3.0	5.0
Netherlands	1 243	1 344	1 494	1 560	1 523	1 346	1 202	1 373	1 196	1 206
Norway	465	499	560	463	734	588	620	617	735	740
Portugal	74	59	54	49	34	73	123	141	168	101
Spain	1 135	1 180	1 456	1 914	1 168	1 593	1 895	1 484	1 065	904
Turkey	215	271	241	304	3 336	496	559	600	512	748
UK	8 189	8 307	9 240	9 881	9 878	9 270	8 859	8 736	7 830	7 141
Total	**29 217**	**30 197**	**32 473**	**33 317**	**33 417**	**33 769**	**35 403**	**35 026**	**33 376**	**30 035**

Source: Deger, S., 'World military expenditure', SIPRI, *SIPRI Yearbook 1991: World Armaments and Disarmament* (Oxford University Press: Oxford, 1991), table 5.10, p. 133.

Appendix D. Exports of major weapons from Western Europe, 1981–90

Figures are values of major conventional weapons systems transferred, in US $m., at constant (1985) prices.

	Supplier												
Recipient	France	FRG	Italy	Spain	Sweden	UK	Belgium Luxembourg Netherlands	Greece Portugal Turkey	Other Europe	Total Europe	USA	Other World[a]	Total
1981–85													
France	–	–	–	–	–	–	–	–	–	–	26	112	138
FRG	81	–	16	–	–	66	–	–	–	163	1 502	2	1 667
Italy	97	110	–	–	–	–	–	–	28	235	759	–	994
Spain	683	60	13	–	–	15	–	–	–	771	624	14	1 409
Sweden	–	3	–	–	–	159	–	–	118	280	176	–	456
UK	1 174	–	–	–	–	–	–	–	1	1 175	1 724	6	2 905
Belgium Luxembourg Netherlands	41	1 210	–	–	–	44	–	–	3	1 298	5 317	–	6 615
Greece Portugal Turkey	182	1 481	110	–	–	243	908	–	491	3 415	2 201	40	5 656
Other Europe	472	–	26	–	289	1 050	44	–	21 780	23 661	3 244	219	27 124
Total Europe	**2 730**	**2 864**	**165**	–	**289**	**1 577**	**952**	–	**22 421**	**30 998**	**15 573**	**393**	**46 964**

USA	37	177	44	–	–	404	12	–	–	674	–	258	932
Other World[a]	16 738	4 659	5 136	1 662	243	5 831	521	93	44 240	79 123	40 821	11 352	131 296
Total 1981–85	**19 505**	**7 700**	**5 345**	**1 662**	**532**	**7 812**	**1 485**	**93**	**66 661**	**110 795**	**56 394**	**12 003**	**179 192**
1986–90													
France	–	–	–	22	–	22	50	–	–	94	487	19	600
FRG	70	–	8	–	–	78	–	–	–	156	2 756	57	2 969
Italy	32	81	–	–	24	–	–	67	28	232	634	–	866
Spain	742	53	136	–	–	11	–	–	75	1 017	4 478	57	5 552
Sweden	76	33	–	10	–	48	–	–	–	167	232	–	399
UK	496	–	–	–	–	–	18	–	4	518	795	178	1 491
Belgium, Luxembourg, Netherlands	57	303	–	–	–	–	–	–	11	371	2 130	9	2 510
Greece, Portugal, Turkey	991	1 458	92	225	–	133	205	–	191	3 295	4 565	336	8 196
Other Europe	216	828	28	–	915	623	59	–	19 775	22 444	1 006	50	23 500
Total Europe	**2 680**	**2 756**	**264**	**257**	**939**	**915**	**332**	**67**	**20 084**	**28 294**	**17 083**	**706**	**46 083**
USA	6	7	3	31	3	316	–	–	112	478	–	558	1 036
Other World[a]	11 111	1 956	1 311	802	859	6 483	1 495	25	45 092	69 134	37 502	12 181	118 817
Total 1986–90	**13 797**	**4 719**	**1 578**	**1 090**	**1 801**	**7 714**	**1 827**	**92**	**65 288**	**97 906**	**54 585**	**13 445**	**165 936**

[a] All countries except Europe and the USA.

Source: SIPRI arms trade data base.

Appendix E. Selections from treaties on the European Community

TREATY ESTABLISHING THE EUROPEAN COMMUNITIES ('TREATY OF ROME')

Excerpts

Article 223

1. The provisions of this Treaty shall not preclude the application of the following rules:

(*a*) No Member State shall be obliged to supply information the disclosure of which it considers contrary to the essential interests of its security;

(*b*) Any Member State may take such measures as it considers necessary for the protection of the essential interests of its security which are connected with the production of or trade in arms, munitions and war material; such measures shall not adversely affect the conditions of competition in the common market regarding products which are not intended for specifically military purposes.

2. During the first year after the entry into force of this Treaty, the Council shall, acting unanimously, draw up a list of products to which the provisions of paragraph 1 (b) shall apply.

3. The Council may, acting unanimously on a proposal from the Commission, make changes in this list.

European Communities, *Treaties Establishing the European Communities* (Office for Official Publications of the European Communities: Luxembourg, 1973), pp. 329–30.

SINGLE EUROPEAN ACT AND FINAL ACT

Title III: Treaty Provisions on European Cooperation in the Sphere of Foreign Policy

Excerpts

Article 30, section 6

a) The High Contracting Parties consider that closer cooperation on questions of European security would contribute in an essential way to the development of a European identity in external policy matters. They are ready to coordinate their positions more closely on the political and economic aspects of security.

b) The High Contracting Parties are determined to maintain the technological and industrial conditions necessary for their security. They shall work to that end both at national level and, where appropriate, within the framework of the competent institutions and bodies.

c) Nothing in this Title shall impede closer cooperation in the field of security between certain of the High Contracting Parties within the framework of the Western European Union or the Atlantic Alliance.

Council of the European Communities, *Single European Act and Final Act* (Office for Official Publications of the European Communities: Luxembourg, 1986), p. 39.

LIST OF PRODUCTS REFERRED TO IN ARTICLE 223 OF THE 'TREATY OF ROME'

Order 255/58 of 12 April 1958

The provisions of Article 223 paragraph 1 b) of the Treaty of Rome are applicable to the arms, munition and war material specified below, including nuclear arms:

1. *Portable and automatic firearms,* such as rifles, carbines, revolvers, pistols, sub-machine guns and machine guns, except for hunting weapons, pistols and other low calibre weapons of the calibre less than 7 mm.

2. *Artillery, and smoke, gas and flame-throwing weapons* such as

(a) cannon, howitzers, mortars, artillery, anti-tank guns, rocket launchers, flame-throwers, recoilless guns
(b) military smoke and gas guns

3. *Ammunition for the weapons at 1 and 2 above*

4. *Bombs, torpedoes, rockets and guided missiles*:

(a) bombs, torpedoes, grenades, including smoke grenades, smoke bombs, rockets, mines, guided missiles, underwater grenades, incendiary bombs
(b) military apparatus and components specially designed for the handling, assembly, dismantling, firing or detection of the articles at (a) above

5. *Military fire control equipment*:

(a) firing computers and guidance systems in infra-red and other night guidance devices
(b) telemeters, position indicators, alti-meters
(c) electronic tracking components, gyro-scopic, optical and acoustic
(d) bomb sights and gun sights, periscopes for the equipment specified in this list

6. *Tanks and specialist fighting vehicles*:

(a) tanks
(b) military type vehicles, armed or armoured, including amphibious vehicles
(c) armoured cars
(d) half-tracked military vehicles
(e) military vehicles with tank bodies
(f) trailers specially designed for the transportation of the ammunition specified at paragraphs 3 and 4

7. *Toxic or radioactive agents*:

(a) toxic, biological or chemical agents and radioactive agents adapted for destructive use in war against persons, animals or crops
(b) military apparatus for the propagation, detection and identification of substances at paragraph (a) above
(c) counter-measures material related to paragraph (a) above

8. *Powders, explosives and liquid or solid propellants*:

(a) powders and liquid or solid propellants specially designed and constructed for use with the material at paragraphs 3, 4 and 7 above
(b) military explosives
(c) incendiary and freezing agents for military use

9. *Warships and their specialist equipment*:

(a) warships of all kinds
(b) equipment specially designed for laying, detecting and sweeping mines
(c) underwater cables

10. *Aircraft and equipment for military use*

11. *Military electronic equipment*

12. *Camera equipment specially designed for military use*

13. *Other equipment and material*

(a) parachutes and parachute fabric
(b) water purification plant specially designed for military use
(c) military command relay electrical equipment

14. *Specialised parts and items of material included in this list insofar as they are of a military nature*

15. *Machines, equipment and items exclusive-ly designed for the study, manufacture, testing and control of arms, munitions and apparatus of an exclusively military nature included in this list*

IEPG, *Towards a Stronger Europe,* A Report by an Independent Study Team Established by Defence Ministers of Nations of the Independent European Programme Group to Make Proposals to Improve the Competitiveness of Europe's Defence Equip-ment Industry ('Vredeling Report'), (IEPG: Brussels, 1987).

Appendix F. Selected documents of the Independent European Programme Group

ACTION PLAN ON A STEPWISE DEVELOPMENT OF A EUROPEAN ARMAMENTS MARKET

Adopted by the Defence Ministers of the IEPG countries at their meeting in Luxembourg on 9 November 1988

1. Introduction

1.1. At their meeting of 22 June 1987, the IEPG Ministers of Defence agreed on a set of long-term objectives in order to improve the competitiveness of European defence industry.

In general they decided

– that efforts towards a stepwise build-up of a European Armaments Market should be made,

– that obstacles restricting border-crossing competition should be removed,

– that contracts should be placed more readily with suppliers in other countries,

– that research activities should provide for the fullest possible exploitation of European resources in talents and funds,

– that LDDI countries should be included in armaments cooperation.

Ministers instructed NADs [National Armament Directors] to draw up an Action Plan for the purpose of a stepwise creation of a European Armaments Market.

The Action Plan should:

– be selective and establish a priority for the recommendations to be put to practice;

– not be confined to the recommendations in the report; alternative proposals made by nations should also be considered;

– make practical proposals for implementation, including an assessment of any organizational or cost-related implications;

– be based on wide consultation with other interested bodies; in particular the views of industry should be set forth through EDIG [European Defence Industry Group] for careful consideration;

– be presented to Ministers in a form and at a time enabling them to take firm decisions at the next meeting.

1.2. The NADs instructed Panel III to submit the draft of an Action Plan. For the treatment of the four main topics mentioned by the Ministers, namely Europe-wide competition, 'juste retour', technology transfer and cooperation in research and technology, Panel III has established *ad hoc* working groups as follows:

– AHWG 1 Competition—chaired by the UK,

– AHWG 2 'Juste Retour'—chaired by NO [Norway],

– AHWG 3 Technology Transfer—chaired by IT [Italy],

– Cooperation in Research and Technology is being handled by SG/6 [Subgroup 6] which has been existing for sometime and is chaired by NL [the Netherlands].

In addition, a SG/7 chaired by GR [Greece] was created. That SG deals with the particular problems of LDDI countries.

The Action Plan is based on the working results of the above mentioned Subgroups and the discussions of NADs and Panel III.

2. General Remarks

2.1. For the creation of a European Armaments Market the opening up of borders for armaments procurements is undoubtedly the subject of central significance. In this process, important national interests are at stake, and re-orientations of heretofore predominantly nationally aligned procurement policies of the member countries may become necessary. Effective competition presupposes the preparedness on the side of the member countries to increasingly accept crossborder purchasing and also admit over time changes in the existing structure of national armaments industries.

The opening up of borders must not be effected one-sidedly, but reciprocally. For industries, approximately similar competitive conditions must be created.

2.2. Because of very important national interests, IEPG countries will only be prepared to admit border crossing competition if they are sure to get an equitable and fair return back in a suitable time corresponding to their vital interests and their possibilities. Therefore some kind of 'Juste Retour' has to be arranged.

There might be a potential conflict between 'Juste Retour' and the call for competition. But 'Juste Retour' is not an end to itself, it is a means to gain the support of the member countries for a border-crossing competition. Therefore and in order to safeguard the possibility to finance an adequate conventional defence, the economy aspect of procurements should as a rule take precedence over perfectionist and mechanistic 100 per cent return.

'Juste Retour' needs to be considered in the context of seeking to secure an efficient and geographically/technologically balanced European defence industry.

In order to provide for cost-effective solutions, the aim is to achieve 'Juste Retour' on a multi-project basis and over a reasonably extended period of time. On the other hand, concrete invitations to bid are not to be interlinked with conditions which might jeopardize competition.

2.3. Unobstructive Technology Transfer is a unique means to improve the efficiency of the IEPG industrial base. If technology transfer is guaranteed, unnecessary parallel work can be avoided and the resources available within the IEPG can be utilized in the best possible way.

The arrangements to be made in this respect should refer to technologies financed by the Governments. IEPG nations should ensure that all appropriate and necessary measures (including the securing of adequate user rights as appropriate) exist so that the intellectual property (proprietary rights, know-how) in Government-funded national and collaborative projects is available for disposition, by the usual commercial processes to facilitate subsequent use by other IEPG nations for defence purposes.

2.4. Besides the border crossing competition the comprehensive and systematic cooperation in research and technology will be the center place for the creation of a European Armaments Market.

After a beginning has been made with CTP [Cooperative Technology Projects] towards cooperation in research and development, it will now be of decisive importance to intensify this process and to concentrate work first and foremost on systematic R&T-cooperation based on technological priority areas. In this context, the question of funding is of considerable importance.

The most effective solution would be common funding as suggested by the EDIG Team. But this was for the present not supported by the IEPG Ministers. In the meantime there is unanimity that common funding is the only long-term solution which will help. This requires major political decisions.

2.5. In realizing the European Armaments Market the special problems of LDDI countries are to be taken into account.

On the one hand the creation of a European Armaments Market will open up to the LDDI countries new and in the past not existing chances. On the other hand because of the special situation of these countries it might be necessary to concede to LDDI countries special transition-periods for the opening of their borders, although such protective measures may also have disadvantages for the favoured industry in view of its future competitiveness.

According to the present knowledge there is a considerable imbalance in armaments relations to the disadvantage of the LDDI countries. Compensating measures by countries having a positive account are necessary.

3. Outline Approach

3.1. In order to achieve border crossing competition i.a. the following steps have to be taken.

3.1.1. To develop measures which will widen the awareness of forthcoming bidding opportunities throughout IEPG countries.

To that effect, a common concept needs to be developed which is to be used to advertise forthcoming bidding opportunities. This concept should include an agreed classification (by nature and by value) of bidding opportunities appropriate to IEPG-wide competitive procurement. It should comprehend as many categories of armament equipment as possible.

The call for a Europe-wide competition covers both national procurements and collaborative projects.

3.1.2. To arrange in each contracting authority focal points to which companies of all IEPG-countries can signify their interest of being known as potential supplier.

3.1.3. To secure compatibility of national procedures for inviting bids and placing contracts. Account should be taken as appropriate of significant variations in price structure.

In this context, it is of particular significance that obstacles in the way of a border-crossing competition be removed and that selection criteria be clarified. As a principle awards should be made on the basis of the most economical solution for the particular requirements.

3.1.4. To develop general guidelines to ensure that in collaborative programmes work-sharing requirements are applied with suitable flexibility so as to promote IEPG-wide competition including countries not participating in the particular programme.

It is to be examined, too, whether guidelines will be required to facilitate the establishment of industrial consortia across national frontiers recognizing their potential role as a means for facilitating juste retour, LDDI involvement and technology transfer.

3.1.5. The development of a standardized reporting procedure to be used to record and report to the IEPG bid opportunities publicized on an IEPG-wide basis and contracts awards implemented on a crossborder basis, including any subcontracts.

This method is expected to provide for a general overview of the extent to which the overall objective, i.e. the establishment of a European Armaments Market, is being attained.

3.2 For the purpose of 'Juste Retour' the following procedure has to be worked out.

3.2.1. Elaboration of a standardized recording system to record all crossborder contracts as a basis for annual reports of the member countries to the IEPG.

3.2.2. Application of a procedure to be used by member countries to annually report to the IEPG crossborder contracts.

To that effect, a common format, the currency unit and the recipient (coordinating agency) must be laid down.

3.2.3. Elaboration of a clearing procedure to summarize national reports into an IEPG-wide report reflecting in condensed form the balance of border-crossing armaments traffic in Europe.

Furthermore, a decision will have to be taken with a view to the establishment of a central coordinating agency. That responsibility may be carried out by a member country as a pilot function or by a Central Registry yet to be established.

3.2.4. The means leading to a correction of possible disparities in the armaments balance have to be evaluated. 'Juste Retour' should be handled in a pragmatic and flexible approach. The countries concerned will have to discuss remedial action on a case-to-case basis.

The country obliged to made a correction, should include the country entitled to it in national procurements and collaborative projects if possible.

3.2.5. At a later time, after sufficient experience has been accumulated, possible improvements of the initially established reporting system should be examined, e.g. the questions of escalation, weighting factor, coverage (scope).

3.2.6. On a longer-term basis, the reporting system should also be amended by contracts awarded by IEPG countries to industries in other NATO countries to provide for an approximate assessment of the IEPG's degree of selfsupport [sic].

3.3 Technology Transfer calls for:

3.3.1. The setting up of a procedure and organizational arrangements for regular exchange of information on existing technological know-how and on current or forthcoming work to investigate opportunities for collaboration and to avoid a duplication of efforts.

3.3.2. The revision and if necessary agreement on general MOUs [Memoranda of Understanding] with the aim of facilitating the exchange of technological information.

3.3.3. Revision of national rules with a view to enabling Governments to extend user rights of the respective research and development result to IEPG Governments.

3.3.4. Revision of existing practices for technology transfer. This includes the scope of payment for and protection of user rights.

3.4. To promote a coordinated R&T-cooperation the following measures are necessary.

3.4.1. Agreement upon technological priority areas on which future cooperation in research and technology is to be concentrated as a matter of precedence including the financial scope of such cooperation.

3.4.2. Longer-term harmonization of national R&T planning/concepts, with a possibility of developing a European Technology Plan (ETP) in the future including the financial scope of such an ETP.

3.4.3. Development of evolutionary funding models for R&T cooperation and for a European technology plan to be established:

– Creation of set-aside funds combined with an adjustment of national budget rules allowing amounts not spent to be carried over from one year to another, in the meantime the setting of target figures combined with precise book-keeping on money spent for CTPs and other R&T cooperation as appropriate.

– Ultimate goal: common funding.

3.4.4. Stronger involvement of industries in the CTP process and in the definition of priority areas. Study of EUREKA and ESPRIT-programs.

Examination of the possibility to establish European research centres in the field of armaments research. What is being thought of are so-called 'co-located research teams' consisting of scientists from several IEPG countries (industry and government establishment) and residing at existing research facilities.

Stronger involvement of the LDDI countries in the CTP process.

IEPG document no. NAD/D-22 (IEPG: Brussels, 23 Sep. 1988).

POLICY DOCUMENT ON THE EUROPEAN DEFENCE EQUIPMENT MARKET

Attached to the Communiqué of the IEPG Ministerial Meeting in Copenhagen on 16 November 1990.

1. In June 1987, the Defence Ministers of the IEPG decided that the defence technological base within the IEPG should be strengthened and that the most effective use of resources should be ensured through the increased harmonisation of operational requirements, co-operation in research and development, and open purchasing arrangements, for defence equipment.

2. In pursuance of the above stated objectives and retaining national responsibility for decision making this Policy Document draws together the principles for the operation of the open European defence equipment market (EDEM), including technology transfer and juste retour, within the IEPG. It has been approved by Defence Ministers as the basis upon which they expect to see, reciprocally, progress being made step by step on this European initiative.

3. The Defence Ministers stress the importance which they attach to defence co-operation among all members of the North Atlantic Alliance and their concern to avoid any measures which might be construed as protectionist. They furthermore recognize that suppliers from non-IEPG countries will be generally free to compete in the open EDEM, consistent with IEPG Governments' policies and any multi- or bilateral arrangements existing between IEPG and non-IEPG countries.

4. Within the framework of paras 1 to 3 above IEPG countries will pursue the commitment to support those countries within the IEPG with developing defence industries (DDI) and will create an open market for defence equipment based on the principles of cross-border competitive purchasing.

5. Such a market is regarded as essential at the present stage for improving the value obtained from national defence equipment budgets, and for ensuring the continued development of the technological base, within the IEPG. The broad strategy will include comprehensive co-operation in research and technology, together with more effective procurement through increased harmonisation of operational requirements and armament orders.

6. An Action Plan as agreed by the Defence Ministers in 1988 setting out the necessary steps is being progressively implemented. Recent activities referring to the key features of the Action Plan relate particularly to:

a. visibility of requirements;

b. arrangements for the exchange of pre-qualification data concerning potential national suppliers;

c. bid-invitations and criteria for contract-award;

d. progressive implementation of juste retour for a transitional period;

e. recording the pattern of cross-border transactions within the IEPG, in order to identify trade imbalances; and

f. promoting co-operation in research and the development of technology and co-ordination in the arrangements for the transfer, use and safeguarding of relevant information.

7. To assist in providing greater visibility of requirements at both prime and sub-contract levels, Contracts Bulletins are being published throughout the IEPG. In general, these bulletins announce all national require-

ments, valued at 1 million ECU or more, for the development, supply and maintenance of military equipment, except for:

a. those already subject to the EC Supplies Directive;

b. nuclear weapons and nuclear propulsion systems;

c. warships (excluding their systems);

d. anti-toxic and radioactive agents;

e. cryptographic equipment.

Additional exemptions will be kept to the minimum (i.e. confined to any emergencies, national security, or other national imperatives which particularly apply in the DDI countries during a phase of transition before those countries fully open their markets), and will be notified to other nations as they arise. As regards collaborative programmes, the arrangements will be discussed by the participating countries on a case by case basis, with the aim of keeping competition at all levels of contracting as open as possible. Likewise the competitive approach will also be encouraged even where publication in the Bulletin is not required.

8. The various contracting authorities will adopt as far as practicable the principle of 'mutual recognition' of potential suppliers on an IEPG-wide basis. A network of 'focal points' established within the IEPG will assist in the exchange between governments of data concerning suppliers and in the provision to potential suppliers of general advice on national contracting practices.

9. The contracting authorities within the IEPG will utilise these Contracts Bulletins and supplier prequalification arrangements as the basis for seeking expressions of interest from suppliers throughout IEPG and inviting bids from all those qualified.

10. Cross-border competition will be encouraged not only at prime contract but also at sub-contract level. In the case of collaborative programmes within the IEPG for the development or production of equipment, cross-border competition will be extended as agreed on a case by case basis by the participating countries, to qualified suppliers from both the participating countries and other IEPG countries (with the pattern of work-sharing being decided progressively on the basis of such competition rather than the financial contributions).

11. The fundamental criterion for the selection of the contractor will be the most economic solution for the particular requirement taking into account in particular considerations of cost (both acquisition and life-cycle), compliance, quality, security of supply. But regard will also be paid to the contribution to the European technological base including juste retour and technology transfer aspects.

12. Each IEPG nation participating in defence research development or procurement activities within the European context will expect an equitable return, or juste retour, in relation to its contribution. This concept of juste retour will operate for a transitional period recognizing the need for safeguarding vital national interests in the context of the development of the open market. In this connection, concrete measures for juste retour will be considered, based upon a suitable reporting and recording system for cross-border transactions established and maintained under the aegis of IEPG Permanent Secretariat.

13. It is the aim to prevent unacceptable imbalances arising, with each IEPG nation undertaking a responsibility to that end in its purchasing decisions. Should such imbalances arise, pragmatic action should be taken to provide juste retour, measured on a multi-project basis and over a reasonable extended time-scale, and implemented in such a way as to assist in pursuing the competitive approach to procurement. Suitable concrete measures might include, for example, individual nations formulating bid-invitations on a basis which will specifically encourage the participation by suppliers at prime contract or sub-contract level from those countries seeking a juste retour adjustment and the submissions by industry of contract proposals which responsively address this aspect. In particular cases the latter objective may be facilitated through international consortia, joint-venture or sub-contract arrangements. In the evaluation of contract proposals in this connection, the general approach would be that, where all other factors are balanced, the bid which caters most effectively for the juste retour considerations should be regarded as representing the best offer. In pursuing juste retour particular regard will be given to preventing unacceptable imbalances to the detriment of DDI countries.

14. Close cooperation between IEPG countries is envisaged as regards the development of technology with the aim of strengthening the European capability and of avoiding

unnecessary duplicative work. The EUCLID programme will form the foundation for achieving this in the field of defence research, and thereafter procedures will apply to encourage co-ordinated planning and funding of defence research among IEPG nations on a continuing basis.

15. Where this cooperation involves the transfer of technology, suitable terms for use will be negotiated and arrangements made to provide protection against mis-use, whether such transfer is handled at the governmental or industrial levels. While any pre-existing ownership of intellectual property rights will be fully respected in this connection, the arrangements made will facilitate where necessary the transfer of technology developed on the basis of Government funding; in other cases this will be done as far as is practicable.

16. In compliance with the aims of IEPG it is intended that the operation of the open defence equipment market within the IEPG should offer appropriate technological and economic benefits to the DDI nations. For a suitable transitional period, other IEPG nations will accordingly give favourable consideration to facilitating the participation of DDI nations in collaborative development and other work. They will also be prepared to provide the DDI nations, at the lowest practicable cost, with that technical data, assistance and training, i.e. transfer of technology, necessary to complement DDI efforts to develop their industrial and technological capabilities and competitiveness and to agree, on an opportunity basis, any other suitable measures to support the DDI nations.

17. The operation of the open defence equipment market within the IEPG will be kept under review in the light of any relevant wider developments and as experience is generally gained. The purpose in mind will be that the continuing evolution of the open market should proceed with all nations adopting cross-border competition within IEPG as a basic procurement principle and, based on increased equipment harmonisation, lead in the direction of a more integrated and less fragmented European defence technology base. In parallel, the on-going restructuring of the armaments industry bill be expected to minimise existing dis-economies.

18. Appropriate action will be taken by each nation case by case to resolve, expeditiously and equitably, any practical difficulties which may be encountered or any disputes which may arise, in the operation of the open market.

19. Guidelines on the contractual procedures and practices for the open market will be prepared as necessary for the use of contracting authorities within the IEPG.

20. There will be close consultation with the defence industry (through the European Defence Industries Group, EDIG). This will apply in particular to the identification of relevant new technologies, to the scope for meeting requirements more economically and facilitating exports to Allies and for improving cross-border industry-to-industry co-operation, and to the continuing development of contractual procedures and practices, within the IEPG.

IEPG document no. MIN/D-14 (IEPG: Copenhagen, 16 Nov. 1990).

Appendix G. Select bibliography

I. Co-operation in arms production and procurement in NATO and Western Europe

Bittleston, M., *Co-operation or Competition? Defence Procurement Options for the 1990s*, Adelhi Papers, no. 250 (Brassey's/International Institute for Strategic Studies: London, 1990).

Blunden, M., 'Collaboration and competition in European procurement: the issue of democratic accountability', *Defence Analysis*, vol. 5, no. 4. (1989), pp. 291–304.

Callaghan, T., *U.S./European Economic Cooperation in Military and Civil Technology* (Center for Strategic and International Studies, Georgetown University: Washington, DC, 1975).

Clarke, M. and Hague, R. (eds), *European Defence Co-operation: America, Britain and NATO*, Fulbright Papers, vol. 7 (Manchester University Press: Manchester, 1990).

Covington, T. G., Brendley K. W. and Chenoweth, M. E., *A Review of European Arms Collaboration and Prospects for its Expansion under the Independent European Program Group*, report no. N-2638-ACQ (RAND Corporation: Santa Monica, Calif., July 1987).

Drown, J. D., Drown, C. F. and Campbell, K., *A Single European Arms Industry?* (Brassey's: London, 1990).

Edmonds, M. (ed.), *International Arms Procurement: New Directions* (Pergamon Press: New York, 1981).

Ferguson, A. (rapporteur), *Report drawn on Behalf of the Political Affairs Committee on Arms Procurement within a Common Industrial Policy and Arms Sales*, document no. 1-455/83, PE 78.344/fin. (European Communities, European Parliament: Strasbourg, 27 June 1983).

Ford, G. (rapporteur), *Report drawn on Behalf of the Political Affairs Committe on Arms Exports of European Countries*, document no. A2-0398/88, PE 118-374/fin., (European Communities, European Parliament: Strasbourg, 22 Feb 1989).

Greenwood, D., 'Constructing the European pillar. Issues and institutions', *NATO-Letter*, vol. 36, no. 3 (1988), pp. 13-17.

Greenwood, D., *Report on a Policy for Promoting Defence and Technical Co-operation among West European Countries, for the Comission of the European Communities*, document no. III-1499/80 (European Communities: Brussels, 1980).

Hagelin, B., 'Multinational weapon projects and the international arms trade', SIPRI, *World Armaments and Disarmament: SIPRI Yearbook 1984* (Taylor & Francis: London, 1984).

Hall, D., 'European cooperation in armaments research and development', *RUSI Journal*, vol. 133, no. 2 (summer 1988), pp. 53–58.

Hartley, K., *NATO Arms Cooperation: A Study in Economics and Politics* (Allen and Unwin: London, 1983).

Hartley, K., 'Public procurement and competitiveness: a community market of military hardware and technology?', *Journal of Common Market Studies*, vol. 25, no. 3 (1987), pp. 237–47.

Independent European Programme Group, *Towards a Stronger Europe*, A Report by an Independent Study Team Established by Defense Ministers of Nations of the Independent European Programme Group to Make Proposals to Improve the Competitiveness of Europe's Defence Equipment Industry ('Vredeling Report'), (IEPG: Brussels, 1987).

Klepsch, E. (rapporteur), Report Drawn on Behalf of the Political Affairs Committee on European Cooperation in Arms Procurement, document no. 83/78, PE 50944/fin. (European Communities, European Parliament: Strasbourg, 8 May 1978); also published as: *Two-Way Street: USA–European Arms Procurement* (Brassey's: London, 1979; and Crane and Russak: New York, 1979).

Moravcsik, A., 'The European armaments industry at the crossroad', *Survival*, vol. 33, no. 1 (1990), pp. 65–85.

Taylor, T., *Defence, Technology and International Integration* (Frances Pinter: London, 1982).

Taylor, T., *European Defence Cooperation*, Royal Institute of International Affairs, Chatham House Papers, no. 24 (Routledge & Kegan Paul: London, 1984).

US Congress, Office of Technology Assessment, *Arming our Allies: Cooperation and Competition in Defence Technology* (US Government Printing Office: Washington, DC, 1990).

Webb, S., *NATO and 1992: Defense Acquisition and Free Markets*, RAND report no. R-3758-FF (RAND Corporation: Santa Monica, Calif., July 1989).

Wulf, H., 'West European cooperation and competition in arms procurement: experiments, problems, prospects', *Arms Control*, vol. 7, no. 2 (1986), pp. 177–96.

II. Arms production

General

Ball, N. and Leitenberg, M. (eds), *The Structure of the Defense Industry: An International Survey* (Croom Helm: London, 1983).

Haglund, D. G. (ed.), *The Defense Industrial Base and the West* (Routledge: London, 1989).

Todd, D., *Defence Industries: A Global Perspective* (Routledge: London, 1988).

Tuomi, H. and Väyrynen, R., *Transnational Corporations, Armaments and Development* (Gower: Aldershot, 1982).

Western Europe

Adam, B. *et al.*, *L'Europe des armes. Trafics et exportations vers le tiers monde* (Groupe de recherche et d'infomation sur la Paix: Brussels, 1989).

Anthony, I., Courades Allebeck, A. and Wulf, H., *West European Arms Production: Structural Changes in the New Political Environment*, A SIPRI Research Report (SIPRI: Stockholm, 1990).

Assembly of the Western European Union, *Armaments Sector of Industry in the Member Countries*, A Report Prepared by the WEU Standing Armaments Committee, document no. 1051 (WEU: Paris, 20 Jan. 1986).

Brzoska, M., Lock, P. and Wulf, H., *Rüstungsproduktion in Westeuropa*, IFSH-Forschungsberichte Heft 15 (Institut für Friedensforschung und Sicherheitspolitik: Hamburg, 1979).

Faltas, S., *Warships and the World Market* (Martinus Nijhoff: The Hague, 1986).

Fontanel, J. and Smith, R. (eds), *L'effort économique de défense* (Ares: Grenoble, 1985).

United Kingdom

Advisory Council on Science and Technology, *Defence R&D: A National Resource* (Her Majesty's Stationery Office: London, 1989).

Freedman, L., 'The case of Westland and the bias to Europe', *International Affairs*, vol. 62, no. 1 (1987), pp. 57–71.

Hartley, K., Hussain, F. and Smith, R., 'The UK defence industrial base', *Political Quarterly*, vol. 58, no. 1 (1987), pp. 62–71.

Levene, P., 'Competition and collaboration: UK defence procurement policy', *RUSI Journal*, vol. 132, no. 2 (1989), pp. 3–5.

Southwood, P., *Disarming Military Industries* (Macmillan Press: Houndsmill, 1991).

Taylor, T. and Hayward, K., *The UK Defense Industrial Base* (Brassey's/Royal United Services Institute: London, 1989).

The Monopolies and Mergers Commission, *The General Electric Company Plc, Siemens AG and The Plessey Company Plc, A Report on the Proposed Merger*, Presented to Parliament by the Secretary of State for Trade and Industry by Command of Her Majesty (Her Majesty's Stationery Office: London, 1989).

Walker, W. and Gummett, P., 'Britain and the European Armaments Market', *International Affairs*, vol. 65, no. 3 (1989), pp. 419–42.

France

Boniface, P. (ed.), 'L'industrie française de défense', *Relations internationales et stratégiques*, vol. 1, no.1 (1991), pp. 7-91.

Chesnais, F. (ed.), *Compétitivité internationale et dépenses militaires* (Économica: Paris, 1990).

Delegation Generale d'Armement (DGA), *L'industrie française de défense* (DGA: Paris, 1990).

Dussauge, P., *L'industrie française de l'armement* (Économica: Paris, 1985).

Hébert, J.-P., *Les ventes d'armes* (Edition Syros: Paris, 1988).

Hébert, J.-P., *Stratégie française et industrie d'armement* (Fondation pour les Etudes de Défense Nationale: Paris, 1991).

Kolodziej, E. A., *The Making and Marketing of Arms: The French Experience and its Implications for the International System* (Princeton University Press: Princeton, N.J., 1987).

Pinatel, J.-B., *L'Economie des forces*, Les sept epées, cahier no. 5 (Fondation pour les Etudes de Défense Nationale: Paris, 1976).

Federal Republic of Germany

Albrecht, U., Lock, P. und Wulf, H., *Mit Rüstung gegen Arbeitslosigkeit?* (Rowohlt Verlag: Reinbek, 1982).

Bontrup, H. J. and Zdrowomyslaw, N., *Die deutsche Rüstungsindustrie* (Distel Verlag: Heilbronn, 1988).

Brzoska, M., *Rüstungsexportpolitik* (Haag und Herchen: Frankfurt, 1986).

Bundeskartellamt, *Beschluß in dem* Verwaltungsverfahren *Daimler-Benz AG u.a.*, document no. B 7-35 00 00-U-137/88 (Federal Cartel Office of the FRG/Bundeskartellamt, 7. Beschlußabteilung: Berlin, 19 Apr. 1989).

Cowen, R. H. E., *Defense Procurement in the Federal Republic of Germany* (Westview Press: Boulder, Colo., 1986).

Huffschmid, J., Voß, W. und Zdrowomyslaw, N., *Neue Rüstung—neue Armut* (Pahl–Rugenstein Verlag: Cologne, 1986).

Köllner, L. und Huck, B. J., (eds), *Abrüstung und Konversion. Politische Voraussetzungen und wirtschaftliche Folgen in der Bundesrepublik Deutschland* (Campus: Frankfurt/Main, 1990)

Wilke, P. und Wulf, H. 'Rüstungsproduktion in der Bundesrepublik. Industrielle Überkapazitäten und staatliche Finanzierungsengpässe', *Politik und Zeitgeschichte*, no. B/2, 11 Jan. 1986.

Wulf, H., 'The West German arms industry and arms exports', *Alternatives*, vol. 13, no. 3, 1987, pp. 319–35.

Italy

Battistelli, F., *Armi, un nuovo modello di sviluppo* (Feltrinelli: Roma 1981).

Gobbo, F. *et al.*, *L'industria italiana degli armamenti*, Revista Militare no. 14 (Centro Militare di Studi Strategici: Rome, 1990).

Jean, C., 'La strategia industriale del sistema difesa', Italian Ministry of Defence, *Conferenza nazionale sull'industria per la difesa* (Italian Ministry of Defence: Rome, 1984).

Jean, C., 'Ristrutturazione e prospettive dell'industria italiana degli armamenti', *L'industria*, vol. 7, no. 2 (1986), pp. 326–36 .

Pianta, M. and Perani, G., *L'industria militare in Italia. Ascesa e delino della produzione di armamenti* (Edizione Assoziate: Roma 1990).

About the authors

Ian Anthony (United Kingdom) is a Researcher on the SIPRI arms transfers and arms production project. He is the editor of the SIPRI volume *Arms Export Regulations* (1991), the author of *The Arms Trade and Medium Powers: Case Studies of India and Pakistan 1947–90* (1991), the SIPRI monograph *The Naval Arms Trade* (1990), and a co-author, with Agnès Courades Allebeck and Herbert Wulf, of the SIPRI Research Report *West European Arms Production: Structural Changes in the New Political Environment* (1990). He is the author of chapters in the *SIPRI Yearbooks 1988, 1989, 1990* and *1991*.

Anthony Bartzokas (Greece) is a Research Fellow at the Greek Institute for International and Strategic Studies. He is currently working on a project on defence industrialization in Greece, Portugal and Turkey. He completed his doctoral degree at the Science Policy Research Unit, University of Sussex, with a thesis entitled 'Military technology transfer and domestic arms production: the case of Greece'.

Harald Bauer (Federal Republic of Germany) is a member of the staff of the Berghof-Stiftung für Konfliktforschung, Berlin. He is currently working on a project on West European security policy with a focus on the role of the European Community and the restructuring of the arms industry. He is the author of *Frankreichs Sicherheitspolitik und das geteilte Deutschland* (1987).

Michael Brzoska (Federal Republic of Germany) is Researcher at the Centre for the Study of Wars, Armaments and Development at the University of Hamburg. He is the co-author of two SIPRI books, *Arms Transfers to the Third World, 1971–85* (1987) and *Arms Production in the Third World* (1986).

Jacques Fontanel (France) is Deputy Director of the Centre de Défense et Securité Internationale de Grenoble, and former Dean of the Faculty of Economic Sciences of the University of Grenoble. He is a member of the editorial boards of *ARES* and *Defence Economics* and a member of the International Advisory Board of the *Journal of Peace Research.* He is the author of *L'Économie des armes* (1984) and *French Arms Industry* (1990), and co-author of *L'effort économique de défense* (1985), *Le Désarmement pour le développement* (1986), *La Vérification du désarmament* (1989) and *Les Principes de Politique Economique* (1991).

Philip Gummett is Senior Lecturer in Government and Technology Policy at the University of Manchester. He is co-editor, with Judith Reppy, of *The Relations Between Defence and Civil Technologies* (1988), and editor of a report for the British Parliamentary Office of Science and Technology, *Future Relations Between Defence and Civil Science and Technology* (1991). He has also written widely on other areas of science and technology policy.

Björn Hagelin (Sweden) is a Research Fellow at the Department of Peace and Conflict Research, Uppsala University. He is the author of *Neutrality and Foreign Military Sales: Military Production and Sales Restrictions in Austria, Finland, Sweden and Switzerland* (1990) and of numerous monographs, articles and research reports, mainly on policy, economic and structural aspects of arms production, arms exports and conversion.

Peter Lock (Federal Republic of Germany) is the Managing Director of the Berghof-Stiftung für Konfliktforschung, Berlin. His books include *Rüstungsprozesse und Weltmarkt-Integration* (1987) and, with Ulrich Albrecht and Herbert Wulf, *Mit Rüstung gegen Arbeitslosigkeit* (1982).

Jordi Molas-Gallart (Spain) is a Researcher at the Science Policy Research Unit, University of Sussex, where he presented a doctoral thesis on Spanish arms production, and has conducted research on European arms development and production programmes. He is also a Research Fellow at the University of Glasgow, where he is working on a project on arms modernization in Spain. He is a former NATO Fellow. He has published articles in *Defence Analysis* and *Cuadernos Economía*, and is the author of two chapters in the book *Missile o Microchips. La Conversión de la Industria Militar a Civil* (forthcoming).

Frederic S. Pearson (USA) is director of the Center for Peace and Conflict Studies and Professor of Political Science at Wayne State University in Detroit. His recent studies of Eurpean arms trade and security policy have appeared in numerous books, including a SIPRI publication, *Arms Transfer Limitations and Third World Security* (1988). He has published articles in *Armed Forces and Society*, *The Journal of Peace Research* and *International Affairs*, among other journals, and is the author of prior books on Netherlands crisis decision making (1981) and on the study of international relations (1991). Along with Michael Brzoska, he is completing a volume on the effects of armament during warfare.

Giulio Perani (Italy) is a Researcher at Archivio Diasarmo, a peace research centre in Rome specializing in the analysis of arms production companies. He has written several reports and articles on arms production and is the co-author, with Mario Pianta, of *L'industria militare in Italia* (1991) and *La speza per armamenti in Italia* (forthcoming).

Mario Pianta (Italy) is an economist at the Italian National Research Council Institute for Studies on Research (CNR–ISRDS). He is the author of *New Technologies Across the Atlantic: US Leadership or European Autonomy* (1988), and co-author, with Giulio Perani, of *L'industria militare in Italia* (1991) and *La speza per armamenti in Italia* (forthcoming). He is also co-author of *La riconversione dell'industria militare* (1990).

Louis Pilandon (France) is Professor of Economics at Blaise Pascal University, Clermont-Ferrand. He is a researcher at the Centre d'etude et de recherche sur le developpement international (CEROI). He is a founding member of the Association française des economistes de défense, and the author of several articles on military expenditure. He is the co-author, with Christian Schmidt and Jacques Aben, of 'Quantitative and causal analysis of military expenditure' (1987) and 'The public defence expenditures in developing areas: an empirical model' (1991).

Judith Reppy (USA) is Director of the Peace Studies Program, Cornell University. She is the co-editor, with Philip Gummett, of *The Relations Between Defence and Civil Technologies* (1988), and is the author of numerous chapters and journal articles in the field of defence economics.

Claude Serfati (France) is a Researcher and Lecturer in economics working in association with the Laboratoire de recerche en économie appliquée–Centre d'etude et de recerche sur l'entreprise multinationale (LAREA–CEREM), University of Paris X (Nanterre campus). He is the author of two chapters on the French arms industry in *Compétitivité internationale et dispense militaires* (1990), and co-author of *The French Arms Industry* (1991).

Trevor Taylor (United Kingdom) is Head of the International Security Programme at the Royal Institute of International Affairs. Until 1990 he taught at Staffordshire Polytechnic, where he became Professor of International Relations in 1988. He is the author of *Defence, Technology and International Integration* (1981), *European Defence Cooperation* (1984), and co-author of *The British Defence Industrial Base: Issues and Options* (1990). He has also written numerous articles and chapters on security and defence industrial issues.

Werner Voß (Federal Republic of Germany) is a Research Assistant at the Wirtschaftsarchiv, University of Bremen, where he is working on the effects of the creation of the European Community Single Market for the arms industries and its consequences for conversion strategies. He is the author of *Reduction in Armaments Production: Lessons to be Drawn from Adjustment to Structural Change in the Shipbuilding Industry in the Federal Republic of Germany* (1989), and of numerous articles on the German defence industry and its conversion possibilities.

William Walker (United Kingdom) is Senior Fellow at the Science Policy Research Unit, University of Sussex. He is co-editor of *Technology and the Future of Europe* (1991), and has written extensively on the defence industry, on industrial performance and on nuclear non-proliferation policy.

Herbert Wulf (Federal Republic of Germany) is Senior Researcher and Leader of the SIPRI project on arms transfers and arms production. He is on leave of absence from the Hamburg University Institute for Peace Research and Security Policy. Prior to his work in peace research, he was Director of the German Volunteer Service in India. He is a co-author, with Ian Anthony and Agnès Courades Allebeck, of the SIPRI Research Report *West European Arms Production: Structural Changes in the New Political Environment* (1990), and the author of *Rüstungsexport aus Deutschland* (1989). He has published books and articles on security policy, arms production, arms transfers, conversion, and disarmament and development. He is the author of chapters in the *SIPRI Yearbooks 1985*, *1990* and *1991*.

Index